Calgary's Fire Department Oct 15th 1918

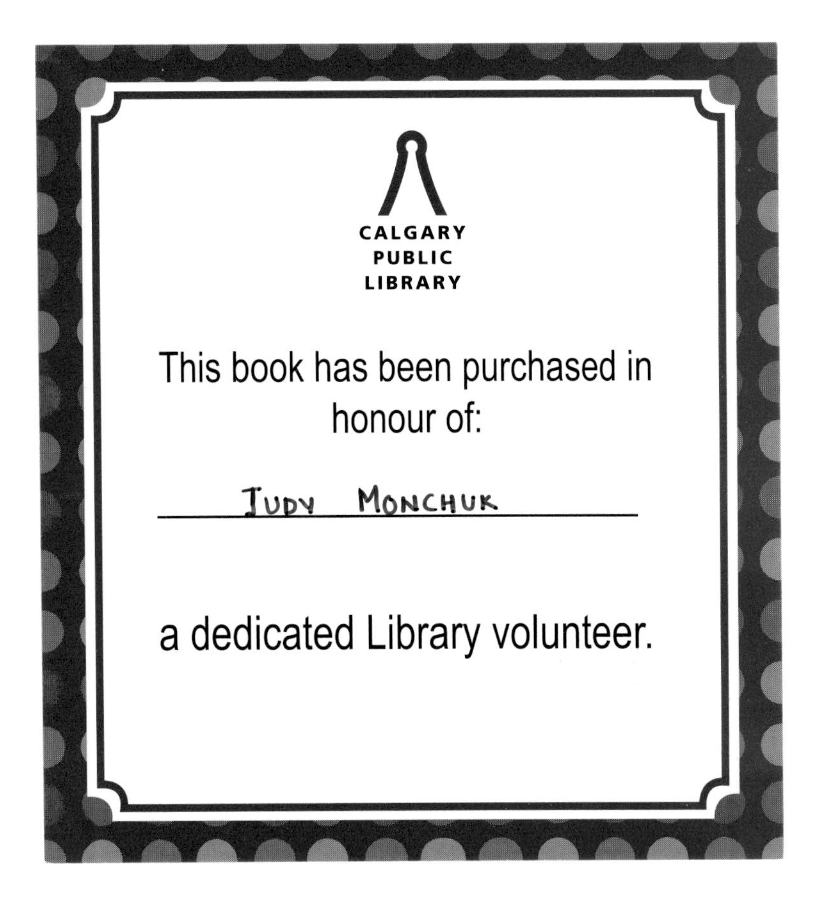

YOURS FOR LIFE
125 YEARS OF COURAGE, COMPASSION AND SERVICE FROM THE CALGARY FIRE DEPARTMENT

MAYOR'S MESSAGE

We rely on those who put their own safety at risk to respond to emergencies.

For 125 years, the Calgary Fire Department has provided that aid and peace of mind, from the volunteers who responded in 1885 to the highly trained professionals who serve us today.

We can rest easy knowing that firefighters are there when disaster strikes. Yet many Calgarians also connect with fire personnel on calmer days through fire prevention programs, community safety initiatives or when they stop by a residential fire hall to have their blood pressure or cholesterol checked.

Quite simply, the dedication of firefighters to public safety makes our city a better place to live. Calgary's history could not be written without noting their substantial contribution.

Congratulations to the men and women of the Calgary Fire Department on this important milestone!

Dave Bronconnier
MAYOR, THE CITY OF CALGARY

▲ Firefighters work to douse the blaze at Central Methodist Church on Feb. 29, 1916.

CHIEF'S MESSAGE

It's been said that one can't go forward without first understanding the past.

The Calgary Fire Department has established a proud tradition of innovation and leadership during its 125 years. Firefighting has changed a lot over that time. There have been vast improvements in apparatus and technology to battle fires. Yet as history evolves, new challenges arise and we will continue to be at the forefront of those battles. Calgary is pushing for national changes to building codes because today's homes are built closer together and materials are more flammable. This means a fire that would destroy one house can now consume three in the same amount of time.

But some things don't change. A firefighter is still the person who runs into a burning building when everyone else is running out.

I hope you enjoy this look through the history of the Calgary Fire Department. Within these pages you will find some of the city's most spectacular fires and a few of the personalities who have left their mark on our department. Some stories will be familiar, while others may be new introductions. A couple may make you smile.

This department has been built by men and women of character who put their lives on the line each day for the citizens of our city.

And that is a legacy we can all be proud of as we move ahead into the next 125 years.

W. Bruce Burrell
FIRE CHIEF, THE CALGARY FIRE DEPARTMENT

Modern firefighter at a Tuscany house fire in 2006. ▶

Many thanks to our sponsors for helping defray costs of this 125TH anniversary book project.

Your generosity will help to keep alive the history of the Calgary Fire Department.

FIRE CHIEF - $40,000
Calgary Fire Department

BATTALION CHIEF - $10,000
Stella Holdings Limited

CAPTAIN - $5,000
ATCO Gas
Bluebird Contracting Services Ltd.
Calgary Stampede

FIREFIGHTER - $2,500
Anonymous
Calgary Fire Fighters Association
Draeger Safety Canada Ltd.
Leaseway Corporation Ltd.
Ledcor Construction Limited
Legacy Fire Protection Inc.
Rocky Mountain Phoenix - Rosenbauer
Safetek Emergency Vehicles Ltd.

FRIENDS - $500
Daniel Gallagher & Darlene Schwab
In memory of Division Chief Lawrence Ashley

◄ Unidentified firefighter in early 20th century gear.

DEFINING VALUES
THE CALGARY FIRE DEPARTMENT CREDO

The Maltese Cross (of St. Florian), with its four arms radiating out from the centre, is recognized around the world as the symbol of firefighters and their willingness to make great sacrifices to protect others. Traditionally, the arms represent courage, bravery, compassion and loyalty to duty, but each fire service personalizes its own badge of honour. Calgary's early firefighters displayed the emblem on hat badges.

The Calgary Fire Department has used its Maltese Cross on uniform shoulder flashes since 1975. The City of Calgary's coat of arms is in the middle, with tools of the trade – a hydrant, ladder and pike pole – on either side.

In 2001, Fire Chief Wayne Morris asked firefighters to select the values they wanted highlighted to represent the Calgary service. Firefighters chose Pride, Professionalism, Respect and Teamwork as guiding principles for the department.

THIS BOOK IS A TRIBUTE TO THE UNSUNG HEROES OF THE CALGARY FIRE DEPARTMENT OVER THE LAST 125 YEARS AND THOSE WHO WILL PROUDLY WEAR ITS UNIFORM IN THE FUTURE.

1885 TO 1913 THE EARLY YEARS

A LEGACY BEGINS

Calgary's early settlers were well aware that on the bone-dry prairie, a wayward ember could wipe out a community. Shouts of "FIRE" might bring well-meaning volunteers running with buckets filled with good intentions and not much else.

IN JANUARY 1885, those intentions did little to extinguish Calgary's first blaze. Bystanders threw snowballs at the flames rapidly destroying the wooden house, a futile attempt to supplement the meagre pails of liquid scooped from the town water tank. The home was reduced to charred rubble in 20 minutes.

By summer, wells had been dug to provide water, the local council agreed to provide some equipment and on August 25, 1885, the Calgary Hook, Ladder and Bucket Corps was formed with 22 willing and able volunteers.

Its first major test was not a roaring success. On a sleepy Sunday morning in November 1886, pealing church bells roused the Fire Department and community as fire quickly spread from hay at the back of a flour-and-feed store to an adjacent warehouse and a tin shop.

▲ Men of the Calgary Hook, Ladder and Bucket Corps in 1886.

1886

CALGARIANS MOVED WHATEVER THEY COULD OUT OF THE PATH OF THE COMMUNITY'S FIRST MAJOR FIRE

RESIDENTS CARRIED FURNITURE, dry goods and sacks of flour away from the rapidly expanding inferno. They were able to remove a keg of gunpowder, preventing a massive blast, but 2,000 rounds of ammunition that remained on store shelves kept exploding as firefighters fought the stubborn blaze. Luckily, no one was shot.

Shifting winds made the fight more difficult and when the damage was tallied, 14 buildings were gone: four hotels, three warehouses, a couple of saloons and a handful of stores.

The brigade was re-organized and additional volunteers brought on board. Calgary purchased a steam engine that could deliver a strong and steady stream of water. A legacy of fighting and preventing fires had begun. ■

1895 SOME OF THE 30 VOLUNTEER FIREFIGHTERS WHO MADE UP THE CALGARY BRIGADE AT THE END OF THE 19TH CENTURY

HORSEPOWER RACES

ONCE THE TOWN PURCHASED A STEAM ENGINE, horses were needed to pull apparatus to fires. Yet it took several years until the brigade had its own stable of horsepower to transport the steam engine, and later, a chemical cart. Until then, the town relied on citizens to grab a team of horses when they heard the fire bell ring, race to the storage house for the engine, hitch the team to the wagon and haul it to the fire. The fastest horseman collected $5 – a sum equivalent to three days' wages at the time.

Unfortunately, there was no rule on whose horses would prevail in a tie.

One day when two horses arrived in a "dead heat," fists were shaken and harsh words exchanged over who would provide the transport. During the argument, a third fire-chaser hitched the engine to his team and made a fast getaway to the fire and his payoff.

The embarrassing incident caused City Council in 1890 to create a payment schedule for equine services: $10 for the first team, $5 for the second and $3 for the third. ■

▲ In the early days, firefighters had to pull equipment to the fire scene. By the 1890s, the Department had its own herd of horses to transport apparatus.

▲ By 1911, the Department had 21 horses in service: 10 matched pairs and the grey mare White Wings, who pulled the one-horse chemical cart and could pair up with other horses.

◄ Horse-drawn combination wagon, circa 1904.

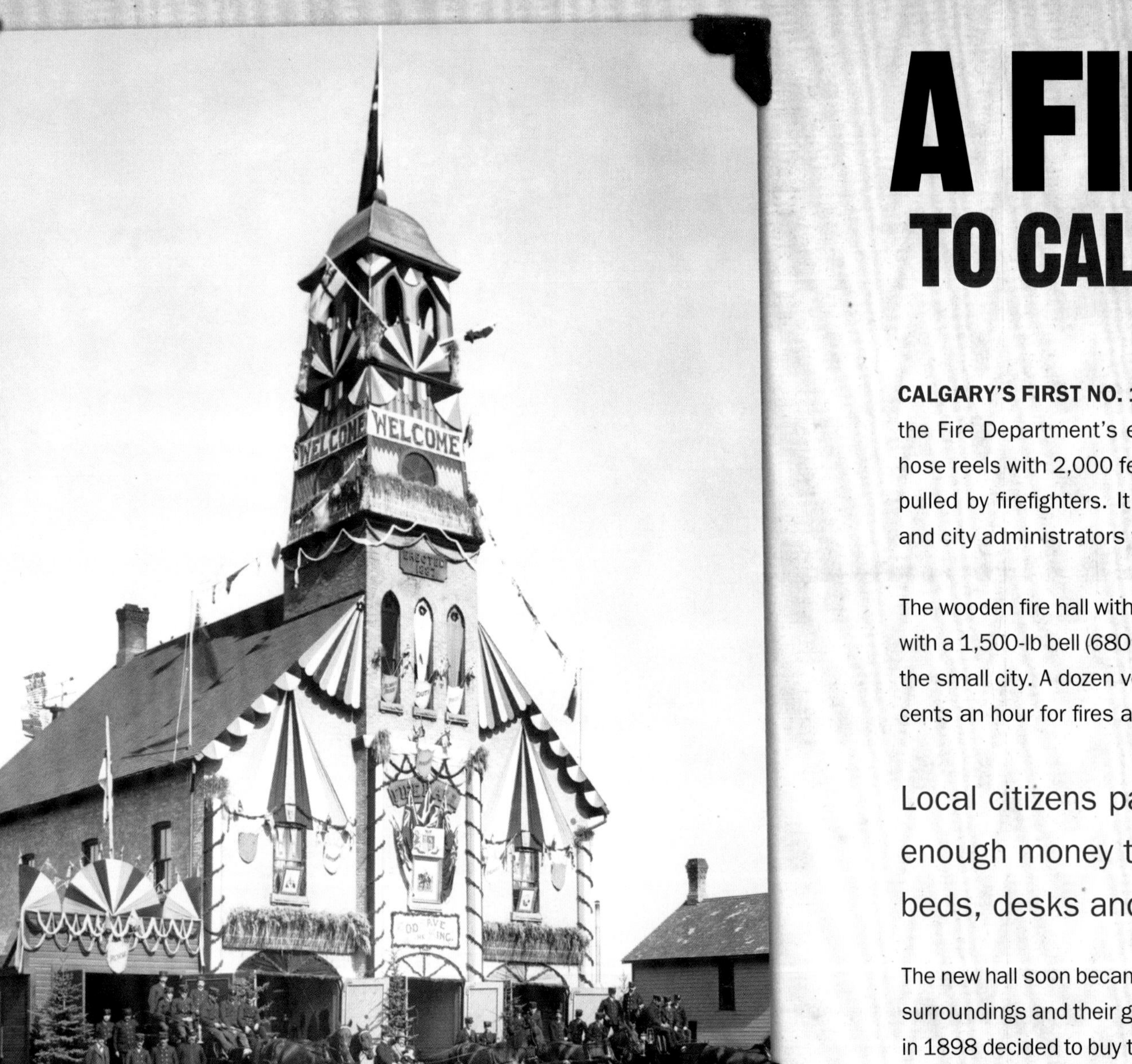

▲ Fire hall No. 1 decorated for the visit of the Duke and Duchess of York.

A FIRE HALL TO CALL HOME

CALGARY'S FIRST NO. 1 FIRE HALL opened in May 1887 to house the Fire Department's equipment: a horse-drawn engine and two hose reels with 2,000 feet of hose (600 metres) that needed to be pulled by firefighters. It was cutting-edge technology for the time and city administrators wanted a building befitting that distinction.

The wooden fire hall with its brick veneer included a church-like spire with a 1,500-lb bell (680 kilograms) – loud enough to be heard across the small city. A dozen volunteers slept in the hall and were paid 50 cents an hour for fires and 30 cents for practice drills.

Local citizens passed a hat and collected enough money to furnish the building with beds, desks and chairs.

The new hall soon became the social centre of Calgary. Proud of their surroundings and their growing stature in the community, firefighters in 1898 decided to buy their own uniforms. The polished outfits cost $18 each. ■

1909 LINEUP OF CALGARY'S FIRST FULLY PAID CREW

▲ Firefighters were expected to have all chores completed, including beds made, by 11 a.m. each day.

▶ Hat badges for Calgary's earliest firefighters included their identification number.

PRIDE IN PROTECTION

LATE IN 1909, the volunteer Fire Department was replaced with a paid brigade of 40 full-time firefighters. For their commitment, the men were paid $70 a month to be available 24 hours a day, with 10 hours off a week.

As the Calgary Fire Department advanced through the next decade, it boasted of having the most advanced firefighting equipment in Canada. In February 1910, Calgary took delivery of a four-cylinder chemical truck that could reach the astonishing rate of 40 miles per hour (64 kilometres per hour) – far faster than the quickest horse.

The truck's speed caught police off guard.

The driver of the high-flying vehicle was pulled over during a training exercise and threatened with a speeding ticket if he didn't slow down the truck. This incensed Chief James "Cappy" Smart, who bluntly told then Police Chief Thomas Mackie that any officer who tried to restrain vehicles en route to a fire was liable to be run down.

City Commissioners hastily drafted a new bylaw allowing the "buzz wagon" to travel to fires as fast as it was able without police harassment.

JAMES "CAPPY" SMART
joined up as part of the Volunteer Brigade in 1886 and became Chief in 1898. Colourful and plain-speaking, "Cappy" Smart transformed the early Department into the modernized envy of his peers and pushed the concept of prevention long before it was fashionable.

▲ In 1910, Calgary became the first community in Western Canada to take possession of motorized apparatus for firefighting. It was quickly dubbed the "buzz wagon" and its crew was known as the "Flying Squadron."

◀ Chief Smart's injuries may have been less serious if he had approved a windshield for his vehicle. But the Chief was worried about the hazard of flying glass.

BUT SPEED ALSO MEANT RISK. In 1912, Cappy Smart's vehicle collided with a streetcar while racing to a fire, severely injuring the Chief and destroying the virtually new motorcar. It took him almost two years to fully recover.

Despite its dangers, the Chief was thrilled with the age of motorization. In a letter to the manufacturer, he noted that the Department's new chemical truck "climbs grades clad with snow when the thermometer is registering 50 below [Fahrenheit], where horses would not go." On the strength of the "buzz wagon's" performance, Calgary ordered three more motorized apparatus: a hose wagon, a combination pump and ladder and a 75-foot (23-metre) aerial ladder truck. ■

▲ Horse-drawn and mechanized apparatus were both key parts of the Calgary Fire Department fleet.

▼ Teams of horses would bring early apparatus to fire scenes.

▼ As time wore on, motorized vehicles became essential to fire operations.

▲ Minnesota-based Waterous Co. was considered the design leader for steam engines that operated pumps for firefighting operations.

1910

CROWDS GATHERED TO WATCH FIREFIGHTERS FIGHT THE BLAZE AT THE CALGARY MILLING COMPANY

1912

FIRE SWEPT THROUGH CHINATOWN, DESTROYING SEVERAL BUILDINGS

PULL IN CASE OF FIRE

ALARM BOXES located across the growing community were the best means of alerting the Fire Department. In 1910, a new system called the Gamewell Gong allowed 56 boxes to be scattered across Calgary, each with a specific code that told dispatch the location of the emergency call.

While the system would be a huge success, it had a design flaw.

Fire logbooks of the day are dotted with false alarms attributed to "Italian posting letter," which appears bigoted to 21st century sensibilities. In fact, the alarms closely resembled mail boxes. Many new immigrants to Canada spoke little English and often didn't realize the emergency nature of the boxes until bells went off. ■

RULES OF THE DAY

THE FIRST REGULATIONS FOR THE CALGARY FIRE DEPARTMENT were adopted in 1909, outlining expectations for everyone from drivers to the fire chief.

Some rules underscored the small size of the Department and Calgary itself, then home to 80,000 people. For example, the Fire Chief was responsible for arranging days off and vacations – a situation that quickly became impractical. And in case of a large fire, the mayor or acting mayor had to be consulted before any building was blown up or torn down.

Yet some rules remain relevant more than a century later. Cigarettes were banned in all stations, as was chit-chat of potentially controversial topics. ■

RELIGION AND POLITICS MUST NOT BE DISCUSSED IN ANY FIRE STATION

– The department is organized and its members are paid to contend with fires, not for any particular set of religious or political opinions.

FROM THE RULES & REGULATIONS OF THE FIRE DEPARTMENT – DECEMBER 1, 1909

SMOKED MEAT

FIRE AT THE BURNS PACKING PLANT

1913

HUGE AMOUNTS OF BEEF WERE PULLED FROM THE RUINS OF THE BURNS ABATTOIR

▲ Firefighters chipped away at the ice formations left by days of battling the Burns fire, where temperatures dipped as low as minus 30 degrees Fahrenheit (-34 Celsius).

Perhaps nothing was as dreaded as a large blaze in the dead of winter, and Calgary's fire history is filled with major battles complicated by weather.

THE 1913 FIRE AT THE BURNS PACKING PLANT was spectacular, raging for seven days in deep-freeze conditions and causing $1 million in damages – a truly breathtaking amount at a time when firefighters were being paid roughly $2.30 a day.

Captain Reg Chambers recalled fighting the blaze for two days and nights. Upon being relieved of firefighting duty, Chambers had to stand in hot water to melt the ice encasing him.

The Calgary Herald noted that poor water pressure made the job of firefighters more difficult, which quickly prompted concerns from a New York underwriter. A letter written to the Fire Commissioner wanted to know when the water pressure would be improved, noting that Calgary's liability insurance was at stake. ■

PACKING PLANT OWNER PATRICK BURNS, one of the Big Four businessmen who started the Calgary Stampede in 1912, was in Toronto on business at the time of the disaster, but never forgot the efforts of firefighters. When Burns died in 1937, a third of his substantial estate was set aside for a memorial trust that aided, among other beneficiaries, the families of fallen firefighters. In 1975, a portion of that fund was earmarked to provide post-secondary education grants to children of firefighters and police officers.

P. BURNS & CO
LIMITED
10

TO 1933

YEARS OF TURMOIL

THE GREAT WAR TO THE GREAT DEPRESSION

Calgary was a boomtown long before the oil-production stampede in the 1970s. As the 20th century dawned, adventurous souls flocked to the young community seeking fortune and a good life.

THE POPULATION JUMPED TO 12,000 FROM 4,100 RESIDENTS BETWEEN 1901 AND 1906. The explosive growth continued and a decade later, Calgary was bursting with 56,500 residents, according to the 1916 Dominion Census. Property values soared. But that influx of people eventually made Calgary more vulnerable to the economic struggles that followed the First World War and eventually, the Depression years.

When Britain called for Canada's help to fight the Germans and other Central Powers in 1914, Calgary was designated a military recruitment hub for Western Canada. Thousands of young men from small towns and farms across the prairies poured into the city to sign up for duty.

► More than half of Calgary's early firefighters signed up to serve in the war.

The patriotic passion burning within many Canadians when war was declared was especially strong at the Calgary Fire Department. More than half of the Department – 54 of 94 members – signed up to serve in the Great War of 1914-1918. Four flags bearing messages of support were sent overseas with the soldiers.

Any Calgary firefighter who volunteered for military service – in the First World War or the Second World War – was deemed to be on a leave of absence. Those hired as replacements had to sign a contract agreeing to surrender the position if the soldier returned and wanted his job. Four of Calgary's firefighters died in the First World War, while others were wounded and never rejoined the department.

Among those who served were brothers Percy and Arthur Simmons, whose extended clan sent 26 members to the front. The family sacrificed two sons, but both Percy and Arthur returned to their firefighting jobs. Arthur died in 1948 fighting a fire at the Union Packing Plant, while Percy retired as a Deputy Chief in 1957 after 45 years of service. ■

FIREFIGHTER TO MAYOR

Three of Calgary's earliest firefighters later moved on to become mayor of the young prairie city.

JOHN W. MITCHELL (Mayor: 1911–13) joined the Calgary Fire Department in 1900. He served five years on City Council and chaired both the Fire and Finance committees before being elected Mayor in 1911. The former firefighter then oversaw construction of the new Fire Headquarters, Station 1, which opened late in 1911. It functioned as a fire station until 1973, when a new Station 1 was built.

W.H. CUSHING (Mayor: 1900–01) was one of the founding members of Calgary's volunteer fire brigade in 1885. A business and political leader, he became Mayor of Calgary in 1900 and later a Minister in the first cabinet of the Alberta legislature.

MICHAEL COPPS COSTELLO (Mayor: 1914–19) joined the Calgary Fire Department in 1898 and was Mayor from 1914 to 1919 – a rare five-year term at a time when many of the city's Mayors only governed a year.

1914

ON NEW YEAR'S DAY, CALGARY FIREFIGHTERS WERE UNABLE TO STOP FIRE FROM DESTROYING THE ENTIRE BUSINESS DISTRICT OF DIDSBURY, ALBERTA.

1915 RINK RAZED

FIRE'S SPEED 'UNEXAMPLED'

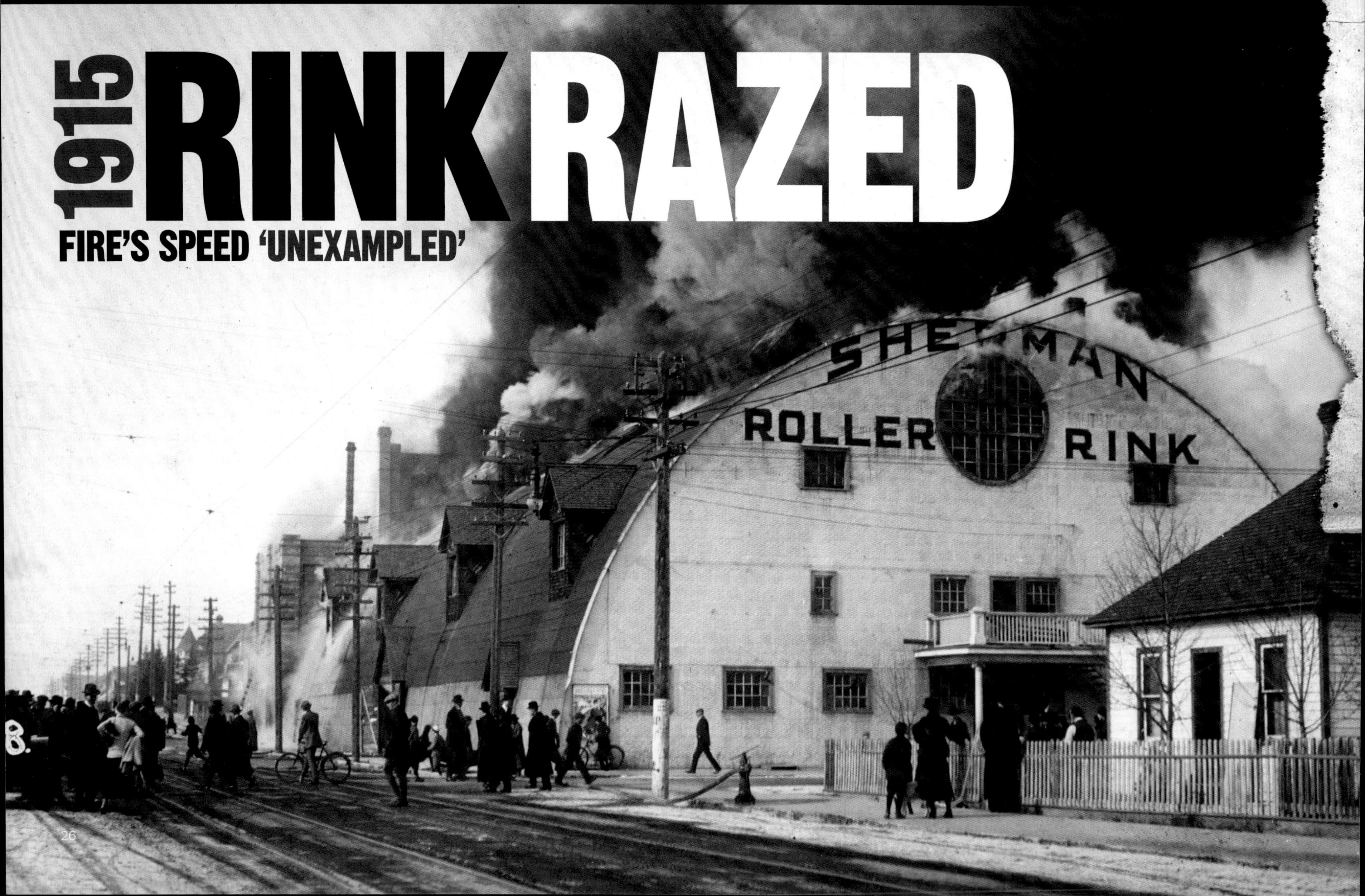

SHERMAN ROLLER RINK

SHERMAN'S RINK TOTALLY DESTROYED BY FIRE AT THE NOON HOUR TODAY

NEIGHBOURING HOUSES AND PLACES OF BUSINESS ENDANGERED BUT FIREMEN SAVED THEM BY WORK

A blazing heap of charred timbers, red hot galvanized iron and shattered concrete blocks was all that remained of the big Sherman rink, Seventeenth avenue and Centre street, at one o'clock today following a fire which originated from unknown causes at 12 o'clock noon.

In exactly one hour the big timber building with the reinforced concrete annex was a total wreck, with eight streams of water pouring on it and raising a vast cloud of smoke and steam.

The buildings, rink proper and the stage annex, were valued at $50,000 and are a total loss.

The rapid progress of the fire was unexampled in the history of the fire department.

◄ A depiction of an article pulled from the January 25, 1915 Calgary Herald.

▼ Firefighters perched precariously on metal supports as they attacked the inferno that destroyed the Sherman rink.

GOING...

GOING...

GONE

FROM THE "CALGARY FIRE DEPT" TO THE 82ND OVERSEAS BATT. C.E.F. MAY 13TH 1916
CHIEF JAMES SMART.
MINNIE SMART. SECY.
SOMME
YPRES
ASST. CHIEF ALEX CARR
GEO. SWEETING
E. GARRATT.
GEO. MURRAY
VILLERS BOIS
H. FISHER.
CAPT N. MACLAUGHIN
VIMY RIDGE
ALBERT
CAPT JUL. SMART.

56th O Bn. C.E.F.

16th March, 1916

Dear Cappy:

We are on the road to Berlin.
Your gift of the flag to this Battalion is lashed to the rear of our Train reminding us of the good friends we have in the Fire Department of the City of Calgary. You can bet your life that if any son of a gun attempts to pull it down we will shoot him full of holes. We are going to endeavour to take this with us to the front at any cost, and we feel perfectly certain that the first German who can read English and sees what is on that flag will turn tail and run.

Again thanking you for the gift to the Boys of the Battalion.

I am, Yours sincerely

William Charles Gordon Armstrong
Lt. Colonel
Officer Commanding, 56th Battalion, Canadian Expeditionary Force

1916 "GIVE 'EM HELL, BOYS!"

Flags of support from those remaining to handle any home fires were much appreciated by troops overseas.

WAR HERO JOINS CFD

THE CALGARY FIRE DEPARTMENT WAS A DESTINATION OF CHOICE FOR MANY MEN – including war hero Raphael Louis (Ray) Zengel, who joined after returning from Europe in 1919. Zengel was awarded the Victoria Cross for his actions in Warvillers, France in August 1918. Zengel, a sergeant, was leading his platoon into attack when German machine-gun fire cut down troops on one side of the line. Zengel raced ahead of the platoon and ran over the artillery platform. He killed the gunner and officer in charge, which caused the rest of the crew to scatter. Zengel's commendation noted that "by his boldness and prompt action, he undoubtedly saved the lives of many of his comrades."

The Victoria Cross is the highest honour that can be awarded to British and Commonwealth forces. Only 94 Canadians and Newfoundlanders were ever honoured with the medal. The American-born Zengel, who was raised in Saskatchewan, was a Calgary firefighter from 1919 until 1927. When he left to take up farming near Rocky Mountain House, Chief James "Cappy" Smart wrote a reference letter praising Zengel's character and expressing regret at his departure. ■

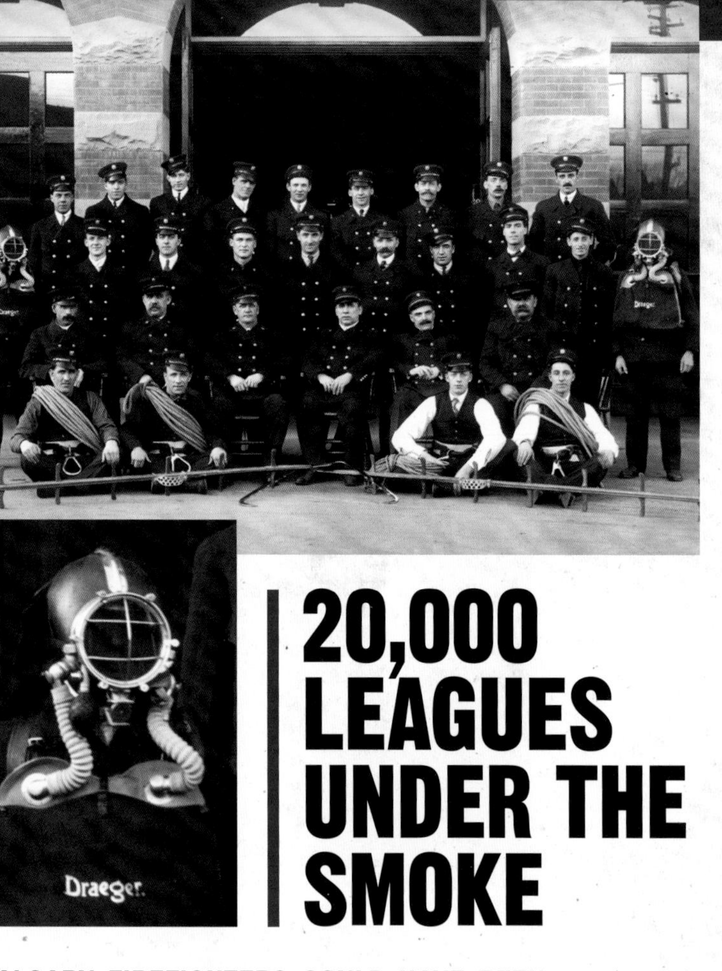

20,000 LEAGUES UNDER THE SMOKE

CALGARY FIREFIGHTERS COULD HAVE BEEN CONFUSED WITH CAPTAIN NEMO'S MEN IN 1914 when the Calgary Fire Department obtained its first breathing apparatus, dubbed "oxygen helmets" by manufacturer Draeger Oxygen Apparatus Co. The devices, which prevented smoke from getting into the eyes and the lungs, resembled underwater diving suits. ■

1920 FOUR PEOPLE DIED WHEN FIRE DESTROYED THE EMPIRE HOTEL

BIGGER & BETTER

MOVING THE CFD FORWARD

► Post-war austerity meant few new vehicles, but in 1928 Calgarians voted to spend $50,000 updating fire equipment. Among the 1929 purchases was the German-made Magirus, an aerial ladder with an 85-foot (26-metre) reach that was dubbed "Maggie." One of only three in Canada, the apparatus allowed for "water tower" firefighting.

▲ Webb aerial ladder, circa 1920s.

▼ Calgary's motorized fleet of firefighting equipment created an impressive lineup outside headquarters.

◄ Chief "Cappy" Smart (second from left) and other members of the Calgary Fire Department posed with the 1926 Studebaker pumper.

▲ The 1922 LaFrance cost $15,000 and could pump 1,000 gallons per minute.

DEPRESSION IMPACT

THE GREAT DEPRESSION HIT WESTERN CANADIAN CITIES WITH ENORMOUS FORCE and Calgary was battered harder than most. Thousands of soldiers had returned to the city after the war ended seeking opportunity, but full-time jobs were in short supply.

After 1929, increasing numbers of people were unable to find work. As the Depression dragged on, municipal officials provided food and shelter for many people in crisis.

By 1934, the City was providing relief to one in eight Calgarians – more than 10,000 people representing thousands of families. Complicating the unemployment problem was the City of Calgary's inability to collect millions of dollars in taxes after the collapse of land prices.

The City's dire finances forced substantial cutbacks at the Fire Department.

Two fire halls were shut down, reducing the number of stations covering the city from 10 to eight. Any positions created by firefighters leaving their jobs went largely unfilled. Yet by 1932, the Department was looking at layoffs. The firefighters' union stepped forward and offered to have everyone work reduced hours to keep comrades on staff – an offer accepted by Chief "Cappy" Smart. Shortly after Cappy's retirement in 1933, three fire halls were closed during the day to further cut costs. ■

AT THE HEIGHT OF THE DEPRESSION, THOUSANDS OF CALGARY FAMILIES WERE RELYING ON RELIEF PROVIDED BY THE CITY TO SURVIVE

FLAMMABLE WHEN WASHED?

MAKING ENDS MEET DURING THE DEPRESSION OFTEN REQUIRED INGENUITY AND CREATIVE THINKING. Unfortunately, desperate times could also lead to dangerous circumstances.

A 1932 fire inspection report details a scenario where a two-storey home was also being used as a laundry. On the ground floor, six large vessels were filled with gasoline, including a washing machine where clothing was soaking in the flammable fuel. Upstairs, sleeping quarters doubled as a drying area for the clothes once the gasoline wash was complete.

In a report to Chief Smart, the fire inspector detailed the hazard:

"This room contains a heater of the radiant type, which is naturally of the open flame type and a direct challenge to human life."

There is no record if tragedy struck the operation before fire officials shut it down. ■

AN ERA ENDS WITH "CAPPY" SMART'S RETIREMENT

In the early years, the Fire Chief directed firefighting operations. He was easily identifiable by his white helmet.

Miniee Smart was the first woman to work for the Calgary Fire Department.

The early days of the Calgary Fire Department were overseen by a larger-than-life leader, James "Cappy" Smart. After working as a carpenter and Calgary's first undertaker, he began his firefighting career as part of the original volunteer bucket brigade in 1885. He became Chief in 1898.

CHIEF SMART WAS A POPULAR PIONEERING SPIRIT, known for his colourful language and hardline stance when it came to fire protection and improving firefighting conditions. His persistence often made Calgary a trailblazer: it was the first Fire Department in Western Canada to get motorized equipment, which quickly paid off in reduced fire losses as crews arrived on scene earlier. In 1918, Chief Smart cited the high cost of horse feed when he asked City officials to consider replacing the Department's horses with a mechanical fleet of vehicles. To reduce the cost, he suggested purchasing eight one-ton chassis of motor vehicles and building bodies for the frames within the fire department's shop. The program had mixed success as a cost-saving measure, but allowed the Department to encourage and showcase innovation.

Among the many "firsts" on Cappy's watch was hiring a woman to work for the Calgary Fire Department: daughter Miniee Smart, who began handling the Chief's office affairs in 1909. Although never known to actually fight fires, Miniee Smart had her own rubber coat and helmet and often accompanied her father to fire scenes. It would be another 80 years before Calgary hired its first female firefighter.

The Fire Chief was considered tough but fair by his men. He repeatedly lobbied on their behalf for better equipment and improved hours. Calgary was the last major city in Canada to grant its firefighters a day off, a move that came after residents approved the practice in a 1933 plebiscite.

Chief Smart was extremely popular with Calgarians, who tended to forgive his rougher edges. Along with his 47 years as a firefighter and 35 years as Fire Chief, Cappy Smart was involved with the Calgary Stampede for 39 years and official timekeeper at all major sporting events in the city for 40 years.

Cappy Smart retired in 1933, but his name lives on with the Fire Department Band that was revived in 1993 and in a local elementary school that was named after him. He died July 25, 1939. ■

1934 TO 1956

CHANGING OF THE GUARD

FIRE GETS NEW SPARK

The decision to grant firefighters one day off a week prompted the first major hiring at the Calgary Fire Department in years and several thousand men applied for the coveted positions.

NOT SURPRISINGLY, THE CLASS OF 1934 WAS AN IMPRESSIVE LOT. Of the 20 new recruits, 15 spent their entire career with the Department, helping introduce and guide change. The group included two future Chiefs: W. Denton (Denny) Craig and Charles Harrison, a star punter with the Regina Roughriders of the young Canadian Football League, who chose to hang up his cleats for a life fighting fires.

Within the ranks was also a future captain of industry, ATCO founder Donald Southern, whose father William had been with the Calgary Fire Department from 1913 to 1929.

Calgary itself was rapidly expanding, steadily annexing land and communities. Its population swelled past 100,000 by the end of the Second World War and to 179,000 in 1956. ■

1940

THE FROZEN REMAINS OF RUTHERFORD HARDWARE STORE

CLASS OF 1934: DON SOUTHERN

FIRE DREAM PAVES WAY TO BUSINESS EMPIRE

DON SOUTHERN VIEWED THE CALGARY FIRE DEPARTMENT AS A WAY OUT OF HARD TIMES. His truck driver's salary had been cut repeatedly as the Depression ground on and he watched lives ground under by financial strife. At one point, with no food for his young family, he cleaned lawnmowers to earn 35 cents – enough money to buy bread, hamburger and some eggs.

A steady job with the Fire Department was a dream come true for the 24-year-old: the monthly salary of $98 provided instant affluence in 1934. Yet by 1941, he was itching for a business of his own and several unsuccessful attempts didn't douse his enthusiasm. In 1946, he saw utility trailers for rent and considered them a good prospect. With wife Ina's savings and a $1,200 bank loan co-signed by a fellow firefighter, Don Southern purchased 15 trailers.

At the time, the Fire Department had an anti-moonlighting rule in effect to ensure opportunities for returning veterans, so son Ron, a 16-year-old high school student, was allotted 40 per cent of the trailer rental business in return for his sweat equity. Early in 1947, Ron Southern registered the operation with the City of Calgary.

The Southerns were soon operating the largest service of its kind in Canada. It was the beginning of what would become the ATCO Group, a global organization of companies with $8.5 billion in assets and 7,800 employees.

Don Southern remained a firefighter until 1951, when he left to focus on his growing business empire, but the links between ATCO and the Calgary Fire Department remain strong today. ■

TOUGH TIMES BREED NEST EGGS

WHILE THE FINANCIAL STRUGGLES OF THE DEPRESSION WERE EASING, banks and mortgage companies were still skittish about extending money for anything less than a rock-solid guarantee.

Firefighters, their jobs inherently dangerous, were not considered acceptable risks by the financial institutions of the day. In 1941, the Calgary Firefighters Credit Union was launched, allowing many firefighters the option of buying their own home.

The Credit Union flourished and grew, eventually providing financial services to all City of Calgary employees. In 1999, its name was changed to Legacy Savings and Credit Union. ■

FIRE & WAR HEROES

THE SECOND WORLD WAR BROUGHT OTHER CHALLENGES. Unlike 1914, Britain's 1939 declaration of war on Germany did not automatically commit Canada to fight, but the country supported the British efforts. As in the initial stages of the First World War, only volunteers served overseas. Yet the City of Calgary was concerned mandatory military service would be invoked if the war dragged on and was less inclined to grant leaves of absence for firefighters than in the past.

In a 1942 letter, Mayor Andrew Davison stressed the difficult situation as he approved leave for one army applicant:

> "We have no desire of preventing any of our employees from enlisting, but we are now facing a serious shortage of competent men in the Fire Department, more especially if all those of military age should be called up in the near future."

Seven senior members of the Calgary Fire Department joined the Corps of Canadian Firefighters to assist British forces in fighting fires caused by German bombs. One firefighter was honoured for his heroism even before leaving Canadian shores.

▲ Calgary firefighter John DeWaal was a hero before he even landed on European soil.

John DeWaal received the British Empire Medal for pulling two soldiers from Ottawa's Rideau Canal during commando manoeuvres. But the bravery was costly: someone stole the firefighter's shoes and wallet carrying his living allowance of $12. In a letter home to his wife, the Calgarian noted it had been quite a day: "I think the Army may pay me back and buy me a new pair of shoes." ■

◄ Hat badge for the Corps of Canadian Firefighters

▲ For the first time, there were too many recruits to fit on one ladder. New firefighters also filled a rig and had to add a second ladder to create the class shot.

CLASS OF 1946

REACHING FOR PATRIOTIC HEIGHTS

AS FIGHTING MEN RETURNED HOME from the Second World War, jobs were at a premium. Many employers, including the Calgary Fire Department, wanted to recognize the sacrifice of those who fought for Canada. When a new class of would-be firefighters was hired in 1946, all 56 recruits were required to have served in the military. The class was brought on board to accommodate the three-platoon system, where firefighters would work an eight-hour day.

The size of the 1946 class was unusual. It would be another 16 years – 1962 – before the Fire Department would hire that many recruits in one year.

Out of the 1946 group emerged two future Fire Chiefs: Derek Jackson, who was Chief from 1972 to 1979, and Frank Archer, Chief from 1979 to 1983. ■

1946

ONE OF THE FEW SHOTS OF THE MAGIRUS AERIAL IN ACTION AS FIREFIGHTERS BATTLE A BLAZE AT THE MCKENZIE BLOCK

1949 FIREFIGHTERS ATTACK THE FIRE AT DAVIDMAN FURNITURE WHILE THEIR COMRADES PREPARE TO CLIMB INTO THE BURNING BUILDING

1949

A FIREFIGHTER RESCUING A CHILD IS AN ICONIC IMAGE, BUT RARELY CAUGHT ON FILM. Calgary Herald photographer Jack DeLorme took this shot of Captain Jack Pilkington carrying a boy from a burning home December 23, 1949 – a picture honoured with Canada's first National Newspaper Award for spot news photography.

▲ Heat doesn't have to be connected with fire. But a worker who succumbed to a hot August day while at the top of Ready Mix Concrete was able to count on firefighters for help.

AWAY FROM THE FLAMES

EXPANDING EXPERTISE

▲ Schoolchildren of all ages took part in poster contests for the Fire Prevention Bureau.

EMERGENCIES AREN'T ALWAYS RELATED TO FIRE and in the 1950s the Calgary Fire Department began to expand its expertise beyond front-line firefighting. This meant increasing rescue capabilities and improving fire education for staff and the public.

The Department revamped its small training program and created Calgary's first fire school in 1955. A three-storey cinder brick building was built for hose drills and other training, while classroom space was obtained at the city's Civil Defence headquarters.

A key part of the firefighting code is the importance of removing hazards long before a spark ignites a blaze. In 1951, Calgary established a Fire Prevention Bureau with five firefighters inspecting buildings for risks. It was part of an increased focus on community protection that would eventually reach into businesses, schools and homes. And it quickly made an impact.

Calgary adopted its first complete Fire Code in 1958 and a year later the city was declared the most "fire safe in Canada" – an accolade based on the number of fatalities, injuries and property damage resulting from fire.

Also in 1958, Calgary became the first city in Western Canada to have a Fire Department aquatic team. Early members used their own equipment and learned rescue and body-recovery techniques in and around the water. Over the years, the team became recognized as a leader in water rescue skills.

In the decades to follow, the Calgary Fire Department would expand its community safety resources. ■

CLASS OF 1936: BILL PHILLIPS

EXTENDING THE REACH OF RESCUE

CAPTAIN WILLIAM (BILL) PHILLIPS literally breathed life into Calgary's public safety program in the 1950s when he was put in charge of the Fire Department's first rescue car.

Phillips, who retired in 1972 as a Division Chief, was credited with introducing mouth-to-mouth resuscitation to the Calgary area after learning the life-saving procedure from a doctor following a fatal accident.

"Most of the other techniques didn't seem to work," he told a newspaper of the day as he offered to teach firefighters and anyone else interested in learning the method.

His efforts attracted the interest of national media and the publicity soon led to controversy. St. John Ambulance refused to review his instructor's certificate because they didn't approve of the technique, which was considered unproven at the time.

Still, the Calgary firefighter persisted in reviving people with artificial respiration and it eventually gained widespread approval by the medical community. In 1967, he was honoured with admittance to the Order of St. John. ■

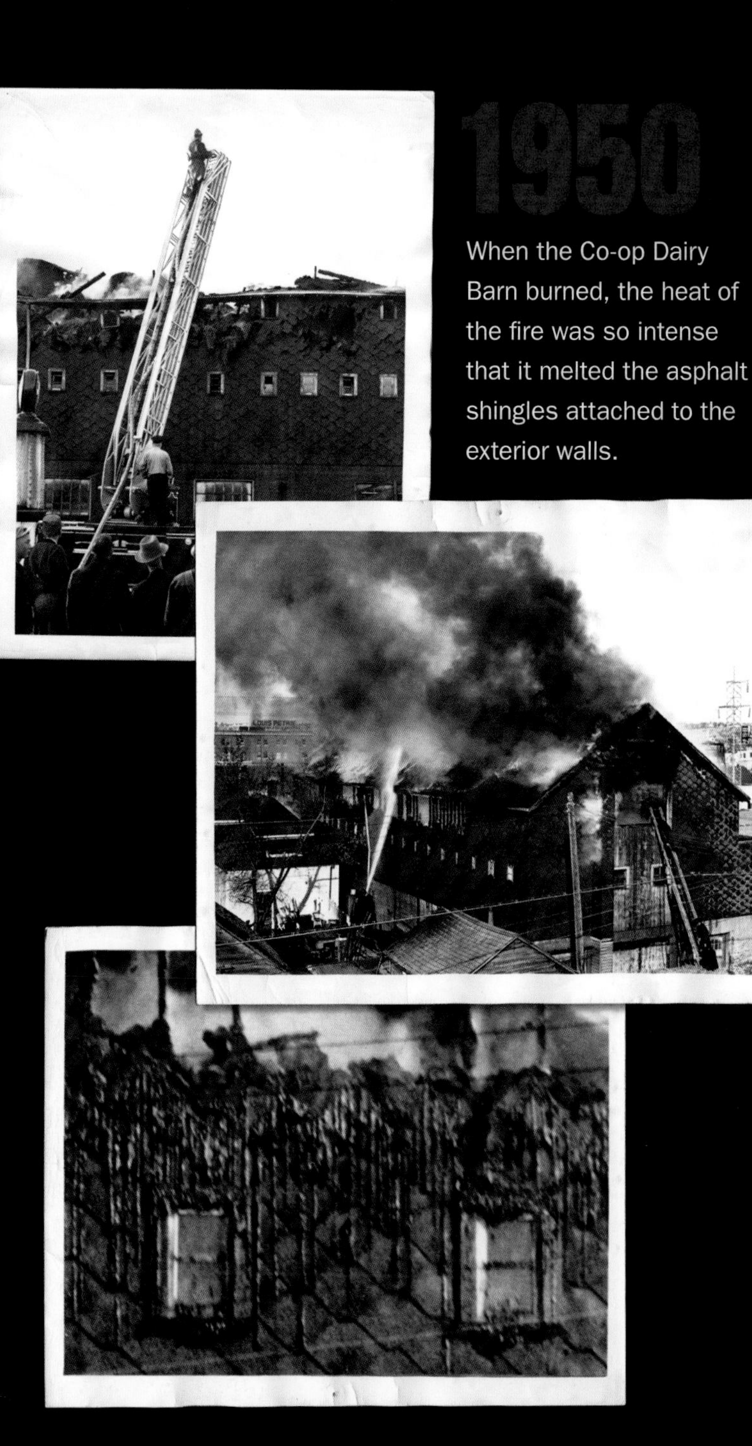

1950

When the Co-op Dairy Barn burned, the heat of the fire was so intense that it melted the asphalt shingles attached to the exterior walls.

1950s DEFENCE TIPS

KEEPING COOL AMID COLD WAR THREATS

COLD WAR POLITICS AND RHETORIC had ordinary citizens worried about the threat of atomic bombs raining down on their communities. In 1954, the federal government prepared a booklet of emergency first aid advice.

Despite the image of a mushroom cloud rising over a crumbling city skyline, the pamphlet from the Civil Defence sector of Health and Welfare Canada kept a light tone. It urged citizens coping with the aftermath of an attack to remain calm until firefighters and other emergency workers arrived. One page reminded people "Don't forget to pass out the cigarettes" to soothe those under stress.

Cartoon illustrations included a man in tattered clothes with a bandaged skull who was playing a banjo to "keep up morale."

Calgary was the site of Operation Lifesaver in 1955, a civil defence evacuation designed to empty a quarter of the city's suburbs within two hours. A CBC documentary of the operation, one of the few exercises done at the time, noted that "War may never come, but if it does, it will be the city with a plan whose citizens will survive." ■

HOSING DOWN FEARS

Firefighters could be counted on to handle burning buildings and whatever else came tumbling down during the Cold War, even if the federal government downplayed dangers to calm the public.

1954

A CHRISTMAS EVE FIRE CA
CONSIDERABLE DAMAGE A
CALGARY PUBLIC MARKET

SACRIFICE OVER STABILITY

BARNEY LEMIEUX

FIREFIGHTING WAS VIEWED AS A STABLE JOB with a decent salary and prestige in the community. But it wasn't always steady work, especially in the early years.

It took some time before Barney Lemieux got the regular paycheque he was seeking when he joined the Department in 1918 after returning from war. He was laid off twice, first in 1921 for 18 months. By 1924, he had enough seniority to retain his job, but offered that position to a colleague with a wife and two children to support.

Yet gaps in service didn't hurt his advancement chances.

The gracious firefighter who stepped aside for a family man was named Fire Chief in 1954, a post he held for more than a decade.

The tenure of Chief Lemieux was marked by significant reorganization of the Calgary Fire Department and a strong focus on training.

"We are going to become the best firefighters in Canada if it kills the lot of you," Chief Lemieux told senior men caught off guard by the tough new standards.

Calgary opened its first fire training school in 1955. ■

WORKING WIVES

FOR DECADES, JOBS WERE SO SCARCE that many organizations and employers, such as The City of Calgary, had an unwritten hiring policy giving preference to family men when positions arose. One firefighter's wife recalled that when the couple married in the 1930s, the City "forbade fire department wives from taking paid positions outside the home." It was felt that wives were taking away jobs.

By the 1950s, this so-called quiet understanding sparked a boisterous backlash from some spouses and Fire Chief George Skene bore the brunt of their anger.

"The women's auxiliary of the day told him to go to hell," recalled Roy Shelley, head of the pensioners' committee in 2008, whose career with the Fire Department began in 1951.

Chief Skene, who oversaw the Department from 1951 to 1954, went to City Council to have the policy changed. ■

1956 FIREFIGHTERS EXAMINE THE ROYAL CANADIAN AIR FORCE EXPEDITOR AFTER THE AIRCRAFT CRASHED AT McCALL FIELD.

In 1958, the Calgary Fire Department was contracted to provide airport emergency services at McCall, later renamed Calgary International Airport. A fire station has been at the airport since 1977.

79

RAPID CHANGE

HISTORY LOST IN HUGE FIRE

Calgary's desire to remain at the forefront of firefighting gained national attention when the city's 1958 Fire Prevention Code was praised by Maclean's magazine as the toughest in Canada.

THOSE RULES, COUPLED WITH ACCOLADES as one of North America's finest Fire Departments, paid off in repeated cuts to fire insurance premiums between 1951 and 1961. But that enviable reputation became difficult to retain as the Department struggled to keep pace with Calgary's rapid growth.

The city's population continued to balloon as the second half of the 20th century progressed. Calgary's population topped 530,000 residents in 1979 – a whopping increase of 350,000 people in less in than 25 years.

▶ When firefighters were unable to save historic Langevin School in 1966, the insurance industry criticized Calgary officials for neglecting fire services while the city grew.

The Calgary Fire Department was stretched to its limits by the protection demands of an ever-expanding territory as the city gobbled up towns and villages including Bowness, Montgomery and Forest Lawn, all of which previously had their own fire departments.

The difficulty of defending such a far-flung area became evident in May 1966 when a spectacular blaze destroyed the Langevin School in northeast Calgary. The school's ruins sparked a fire of controversy when Deputy Chief Denny Craig told City officials the historic three-story sandstone structure built in 1909 could have been saved if the Department had more firefighters.

A more ominous warning was delivered to City Council in September 1966, when the Canadian Underwriters' Association reported that Calgary's rapid expansion had left the Fire Department woefully under-equipped and understaffed.

Upgrading protective services was put on the urgent to-do list and Chief Charlie Harrison was asked to draft a three-year capital plan to cope with the Fire Department's growth needs. To a city known for its boom and bust cycles, the insurance industry's rebuke was a blunt reminder of the need to plan for a growing metropolis during good times. ■

ONE MAN DIED IN THIS GASOLINE TANKER ACCIDENT NORTHEAST OF CALGARY

1959

▲ Designed by Calgary firefighters, the new truck being shown off in June 1956 was capable of shooting water from seven lines and considered the most versatile fire truck in North America.

SWEET MEMORIES

HOLIDAY RITUAL SPARKED BY FIRE

FIREFIGHTERS WHO EXTINGUISHED A KITCHEN FIRE for a wealthy society matron in 1907 had no idea it would kindle a holiday tradition.

It was just prior to Christmas when Helen Cross, wife of rancher and Big 4 businessman A.E. Cross, smelled smoke as she entered her lean-to kitchen. A maid was quickly dispatched to sound the alarm and firefighters quickly had the situation under control.

For the next 50 years, a frosted cake was delivered on Christmas Day to the No. 3 Station – a gift from the Cross family and often baked by Mrs. Cross. When the cake arrived in 1957, one firefighter commented that the tradition had been going on for half a century. Early in 1958, firefighters called on Helen Cross with a bouquet of flowers to thank her for the years of thoughtfulness. ■

1961

TRAGIC RECORD

SEVEN DIE IN APARTMENT FIRE

CALGARY MARKED A SOMBRE MILESTONE IN 1961 WITH THE DEADLIEST FIRE IN THE CITY'S HISTORY

SEVEN PEOPLE DIED ON NOVEMBER 29 when the three-storey Cameron Block erupted in flames before dawn. The five-alarm fire sent residents fleeing into the night, forced to endure sub-zero temperatures clad only in pyjamas.

One man desperate to escape jumped through a window dressed only in his underwear and became frozen to a small marquee sign, his hands and feet fused to the metal. Firefighters freed him from his icy trap, but could not save his life.

Three bodies were later dug out of debris at the bottom of the stairway.

Fire Inspector Logan Brown, who retired as Deputy Fire Marshal in 1975, arrived at the scene as firefighters battled the stubborn blaze.

The Cameron Block was within Brown's regular coverage area and he knew one of the dead: a young woman who had been living in a rougher neighbourhood, but eventually heeded the scolding of the fire inspector and moved to a building where she believed she would be out of harm's way.

"It was a helluva lot safer than the other place," Brown said from his home in Chilliwack, B.C. "It wasn't a firetrap. I thought she'd be safer in the Cameron Block – especially in a suite next to the fire escape."

The devastating fire was later attributed to careless smoking.

Earlier in the century, the Cameron Block was home to the notorious Eye Opener, a muckraking newspaper run by Bob Edwards and known across Canada for its stinging observations and razor-sharp criticism. From 1904 to 1922, the paper skewered politicians and generally mocked authorities of the day. ■

▶ Fire caused $1.7 million in damage to the posh Glencoe Club.

▼ An aerial ladder extended high above the three-storey Haultain School while battling a fire that caused $100,000 damage. Built in 1894 as South Ward School, the sandstone structure was Calgary's first school with electricity and running water.

▲ Flames highlighted the Lone Pine name as the supper club burned late into the summer night.

FIRE & ICE: A WINTER NIGHTMARE

TWENTY-EIGHT FIREFIGHTERS were treated for frostbite after fighting a wicked fire that destroyed a downtown city block during a record cold snap.

Firefighters coped with -30°C temperatures, frozen water lines, intense heat and thick black smoke when fire wiped out the Condon Block in the early hours of December 16, 1964. Calgary's worst cold snap in 40 years complicated the battle, as winds gusting up to 65 kilometres an hour allowed flames to consume business after business: a bank, three stores and a restaurant. ■

1966 A SPECTACULAR FIRE AT THE MICHAEL BUILDING, ORIGINAL HOME OF CALGARY LANDMARK HY'S STEAK HOUSE, CAUSED $600,000 IN DAMAGE

IN PRAISE OF BETTER EQUIPMENT

A FIREFIGHTER'S HELMET is designed for protection from falling debris, but few ever test it in a crisis.

Jim Keats lived to praise his headgear when fire swept through Hy's Steak House on January 21, 1966. By the time crews arrived, terrified occupants were trying to escape the upper reaches of the Michael Building above the venerable restaurant. Some had already jumped. Flames could be seen on the backside of one man leaning out a second-storey window.

The scene was unlike anything the young firefighter had experienced in his seven years on the job. Patricia Keats recalled that her husband described the fire as "a scene out of hell." It triggered an adrenaline rush that allowed Keats and another firefighter to put up an extension ladder where normally four men would have been needed.

As firefighters battled the blaze inside the building, the ceiling caved in and a beam collapsed on Keats and another man, knocking the pair down a landing. Dazed and groggy, the two managed to make their way outside where Keats realized his hard plastic helmet had split in half. "It was a good thing they hadn't heard the beam falling because if they had looked up, they both would have been killed," said Patricia Keats.

Although the Fire Department refused to allow Keats to keep the broken helmet as a souvenir, Fire Chief Charlie Harrison later presented him with a silver-plated version appropriately described as a Lifesaver Award. ■

REMOVING FIRE BOXES CALLING 911

IN 1970, Calgary became one of the first cities in Canada to adopt the 911 emergency system.

Emergency calls were now funneled through the Calgary Fire Department. It was the first step towards eliminating the 450 fire alarm call boxes that had been scattered across the city since 1901.

At the turn of the century, few people had telephones in their homes – a situation that had dramatically changed by 1978. Meanwhile, the Fire Department was becoming frustrated with the high ratio of false alarms received from call boxes: almost 90 per cent over a three-year period.

"Men and equipment are dispatched nearly 1,000 times a year to non-existent emergencies," Deputy Fire Chief of Services Tom Minhinnett said at the time.

"This situation not only endangers the lives of firefighters by putting them on the street unnecessarily, but it also means our firefighters are not available to respond to a real emergency should one occur."

The last box alarms were removed from Calgary streets at the end of 1978. ■

AMBULANCE STRIKE TRIGGERS CHANGE

FIRE TRANSPORTS INJURED

PRIOR TO 1971, Calgary's ambulance services were private companies competing for patients to make a profit.

Operational standards to handle the sick and injured were minimal: a first aid certificate and a clean driver's abstract. But as the 1960s ended, there were demands for better training and higher pay, while doctors wanted patients to receive some stabilization before arriving at hospital.

The situation came to a head with an ambulance strike during the 1970 Calgary Stampede, the city's showcase to the world. A public outcry forced civic officials to act quickly. Within months, the Calgary Fire Department was overseeing an ambulance division that began operating January 31, 1971 with 29 employees.

A decision was also made to increase training to a level being pioneered in the United States. This led to development of a pre-hospital care program at the Southern Alberta Institute for Technology. SAIT's first Emergency Medical Technician program began in 1972 filled with Calgary staff.

The paramedic ambulance service operated out of Calgary fire halls and was unique among Canada's major cities. By 1984, the ambulance division was replaced with the creation of Calgary's Emergency Medical Services Department. ■

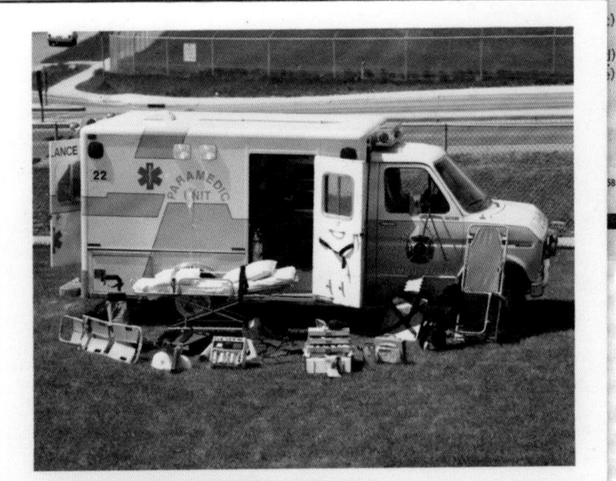

THE CITY OF CALGARY

PUBLIC NOTICE

The City of Calgary Fire Department will commence an Ambulance Service for the Public

Effective Midnight
Sunday, Jan. 31, 1971

To Call An Ambulance
Emergency Dial 911
Normal Calls Dial 261-4000

Ambulance Rates
Within City Limits: $20.00
Outside City Limits:

Emergency: $20.00 plus $1.00 per patient mile outside city limits
Hospital and Nursing Home Rates: By Contract Arrangements

Payment for Ambulance Service:

(a) You will be invoiced next day
(b) Self-addressed envelope will be enclosed with invoice for mailing your remittance to Chief Cashier
) Remittance may also be taken in person to the Chief Cashier, 323 - 7th Avenue S.E.
) You will receive an "Official Receipt" for your payment
) For Ambulance Business Office and to make

ACCOUNT INQUIRIES: DIAL 287-1150

The Board of Commissioners

▲ The Calgary Fire Department ambulance division began in 1971.

ON GUARD FOR OUR FALLEN COMRADES

In 1971 the Calgary Fire Department formed its own Honour Guard, ensuring a formal way to pay tribute to firefighters killed in the protection of others.

THE GUARD, made up of active duty firefighters, was established after the funeral of Lieutenant Harold Smith, who died fighting a huge downtown fire at the McTavish Block on May 27, 1971.

Members who had military experience presented the idea of forming an honour guard, which could provide ceremonial escort at funerals and other official events. Its first appearance was at the National Convention of Fire Chiefs in August 1971. Since then, the Honour Guard has represented the Calgary Fire Department at recruit graduations, retiree banquets and royal visits.

The formal uniform of the Honour Guard includes a sleeve patch with a cluster of red stars: one star for each Calgary firefighter who has died in the line of duty. The Honour Guard also represents the department at sombre occasions such as line-of-duty deaths and memorials for emergency responders across North America. ■

1972

Firefighters battled a blaze at the Beachcomber polynesian restaurant, where firefighter Jerry Walter died.

Crowds gathered to watch a spectacular late-night fire at McArthur Furniture that caused $500,000 in damage.

ATCO praised firefighters for ensuring a $1-million fire did not spread to 30 other buildings in its Lincoln Park complex.

SPOTTING OUR LITTLEST ARSONIST

In 1973, Fire teamed up with Calgary Police to form a joint arson squad. Two police detectives were assigned to help with any requests from fire investigators. The squad examined all suspicious fires, explosions, multiple-alarm fires and any blaze where a death occurred.

Joe Lewko was a Fire Inspector for 28 years and part of the joint arson unit for five years. He was often called to testify in court and proudly notes that no case where he gave evidence was ever lost by the Crown. A no-nonsense type, Lewko heard lots of bizarre explanations over the years for how a fire may have started. One of the oddest was the apartment dweller who blamed a mouse nibbling on a wooden match. The man even offered up a rodent with a blackened nose as proof.

"I told him he wasn't telling the truth," said Lewko, rolling his eyes and laughing at the memory. "If it was the mouse, his whiskers would have been burned off. I said `I bet it was cigarettes' and sure enough, when we pulled the mattress away from the wall, there were cigarette butts."

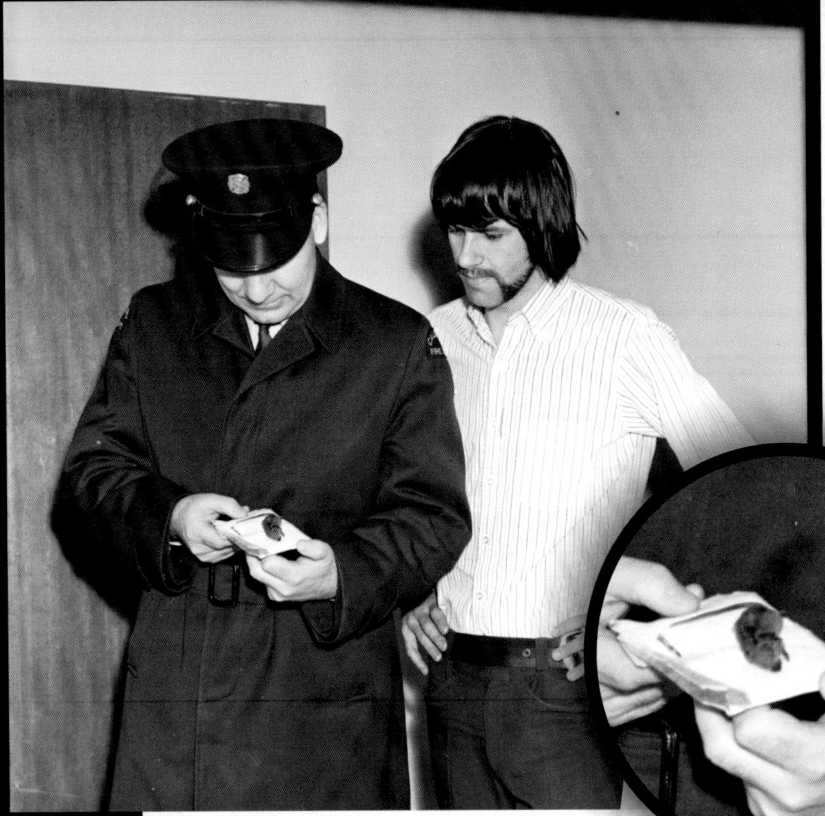

▲ Strange as it might seem, mice have been accused of being mini fire-starters for more than a century. Early logbooks in Calgary have several entries blaming mice.

AS THE 1970s PROGRESSED, THE CALGARY FIRE DEPARTMENT KEPT ADDING AND EXPANDING THE EMERGENCY SKILL SETS NEEDED TO SERVE AN EVER-EVOLVING CITY

CALGARY WAS NOT ONLY GROWING OUT, IT WAS ALSO GROWING UP

More high-rise buildings were built in the booming city, creating a greater need for improved high-angle rescue proficiency.

In 1977, the Department added an elevating platform to its fleet: the Firebird 150. The aerial platform truck was the largest fire vehicle of its day and one of only two in Canada. It could hoist a firefighter 150 feet (46 metres) into the air and had the potential to rescue victims trapped as high as 10 floors above ground.

The Firebird was easily identifiable by its new lime-yellow paint job, part of the Department's effort to improve visibility with a new image. The colour experiment continued until 1991, when City officials agreed with a National Fire Protection Agency report that concluded the traditional red engines were easier to see. ■

Window washer rescue at the IBM high-rise on May 30, 1969.

An aerial platform truck known as the Firebird allowed firefighters to pour water on blazes from up to 150 feet (46 metres) in the air and provided increased rescue capability.

1977

CABINETMAKER SUNGOLD SUFFERED $1.9 MILLION IN DAMAGE. The costs would have been higher if the facility hadn't been located next door to Fire Station 14.

1980 TO 1999

BOOM CITY GOES BUST

DOING MORE WITH LESS

The 1980s brought tough times to Calgary and kept firefighters hopping.

THE INTRODUCTION OF THE NATIONAL ENERGY PROGRAM (NEP) in 1980 ushered in an economic downturn and created among Albertans an enduring suspicion of federal motives. Prime Minister Pierre Elliott Trudeau's government imposed federal authority over energy resources and set new price regimes without consent from Alberta and other oil-producing provinces. The impact of the NEP, combined with a global energy slump, was swift.

By 1982, dozens of energy companies had left Calgary, triggering massive layoffs that began within the industry and quickly spread to service sectors and beyond. Thousands of people lost their homes as personal and business bankruptcies soared.

One of the first signs of a troubled local economy is a jump in deliberately set fires, as cash-strapped individuals look for ways to avoid payments they can no longer afford.

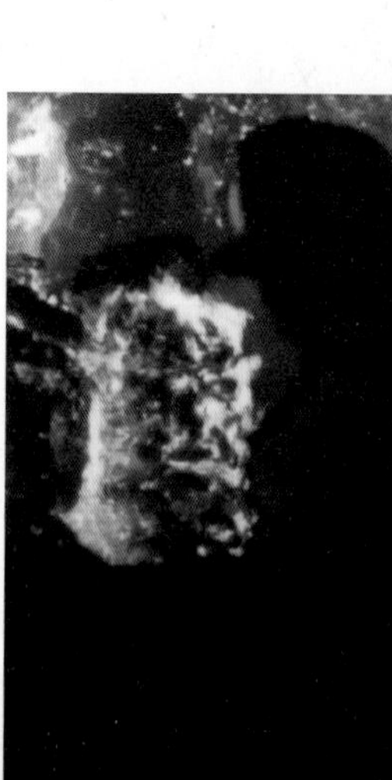

▲ Firefighters were kept busy with a vast array of flaming dangers. Petro Canada even added a helicopter platform to its skyscraper roof in case of fires near the top of its 52 floors.

NEW RECRUITS SAVED BY WAGE FREEZE

JOE LEWKO, the Calgary Fire Department's first official photographer, spent 36 years taking pictures of fires and the after-blaze rubble that could be used as investigation evidence. As a Fire Inspector, he was alarmed by the increase in vehicle fires during the 1980s and developed an expertise in their investigations.

"We were having them almost every week," says Lewko, who wrote Fire on Wheels, the first manual on investigating vehicle fires, after his retirement in 1987. "I was wondering if they were accidental or deliberately set."

Firefighters also coped with roughly 1,100 house fires in three consecutive years: 1980, 1981 and 1982. Although Calgary's population would almost double over the next 25 years, the number of home fires never approached those of the early 1980s. In fact, from 1989 to 2006, there were never more than 600 residential fires in a year.

In 1983, City officials warned that budgets could be cut by as much as 10 per cent and sent pink slips to the final class of recruits hired in 1982. But after a special union vote, firefighters agreed to forgo a uniform allowance to retain the probationary firefighters. The Calgary Firefighters Association struck a deal with the City freezing wages for two years in exchange for assurances of no layoffs.

Fiscal pressures didn't ease up in the 1990s. In 1993, Alberta Premier Ralph Klein ordered sharp rollbacks in government spending to get the province's finances in order and eliminate deficit spending by 1997. The City of Calgary received less money for municipal services, which translated into fewer dollars for Fire and other protective services.

"The motto of the whole era was 'doing more with less'," recalled Fire Chief Jack Ross, who ran the Department from 1988 until 1999.

The number of firefighters remained virtually unchanged between 1982 and 1994 and actually declined at the end of the 1990s.

"We endeavoured to come up with innovative ways to keep things going," the retired Chief said in 2009. "There were a million little things. We tried to adjust our system, but mostly it was trying to keep the team together."

In April 1995, the Calgary Fire Department announced it would hire up to 20 new firefighters – the first recruits in three years. Austerity measures also applied to equipment. In 1998, Calgary had 37 fire engines, the workhorses of the fleet. That was just two more than the Department had 15 years earlier in 1983.

Chief Ross knew that by not keeping pace with city growth, a service backlog was being created that would take years to rectify: "The financial resources just weren't there. We weren't building stations we thought we should."

Despite constraints, the Department remained focused on expanding its expertise. Disaster Services began in the 1980s as a unit within fire, while the heavy rescue team was formed in 1991 to deal with collapsed structures. ■

A series of explosions at the Hub Oil refinery in 1999 killed two men and triggered massive fireballs. ►

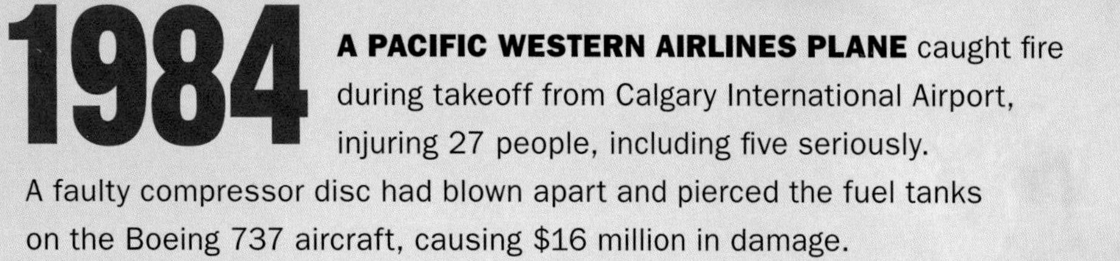

1984 **A PACIFIC WESTERN AIRLINES PLANE** caught fire during takeoff from Calgary International Airport, injuring 27 people, including five seriously. A faulty compressor disc had blown apart and pierced the fuel tanks on the Boeing 737 aircraft, causing $16 million in damage.

$
PAY THROUGH THE YEARS
125 YEARS OF WORKING HOURS & WAGES
A HISTORIC BREAKDOWN OF MONTHLY PAY FOR CALGARY FIREFIGHTERS:

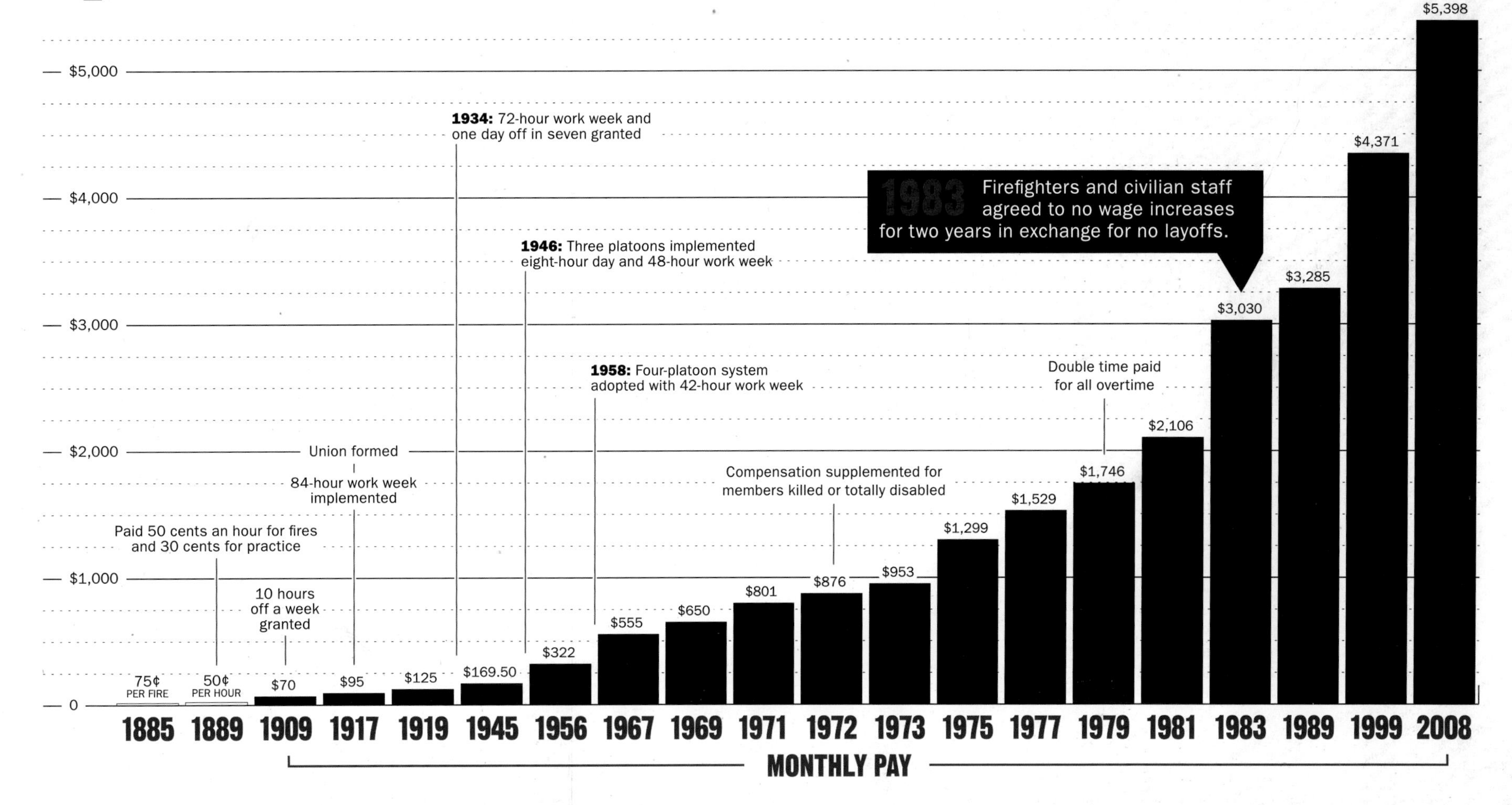
$5,000
$4,000
$3,000
$2,000
$1,000
0
1934: 72-hour work week and one day off in seven granted
1983 Firefighters and civilian staff agreed to no wage increases for two years in exchange for no layoffs.
1946: Three platoons implemented eight-hour day and 48-hour work week
1958: Four-platoon system adopted with 42-hour work week
Double time paid for all overtime
Union formed
84-hour work week implemented
Compensation supplemented for members killed or totally disabled
Paid 50 cents an hour for fires and 30 cents for practice
10 hours off a week granted
75¢ PER FIRE
50¢ PER HOUR
$70
$95
$125
$169.50
$322
$555
$650
$801
$876
$953
$1,299
$1,529
$1,746
$2,106
$3,030
$3,285
$4,371
$5,398
1885
1889
1909
1917
1919
1945
1956
1967
1969
1971
1972
1973
1975
1977
1979
1981
1983
1989
1999
2008
MONTHLY PAY

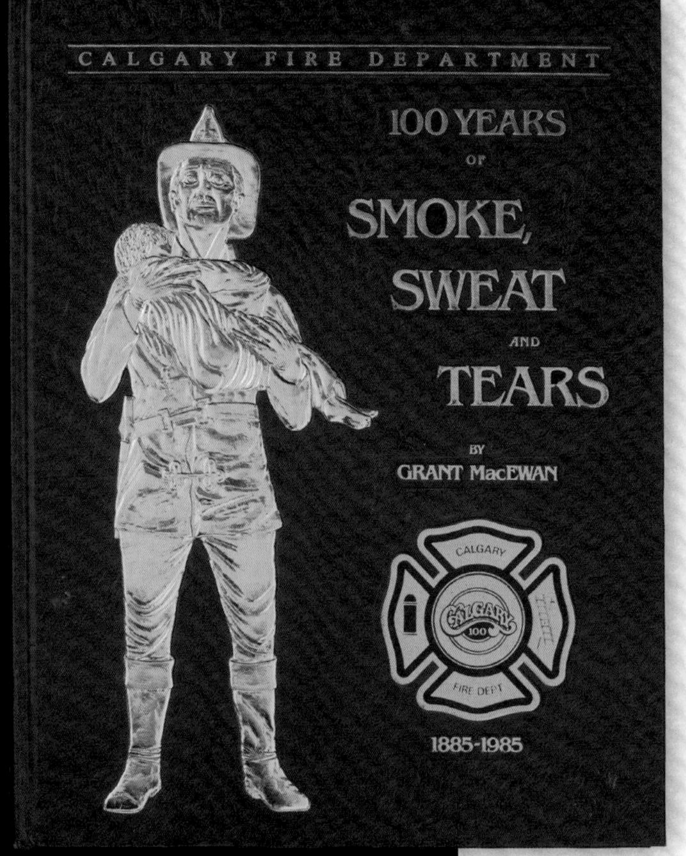

▲ The centennial book became a treasured keepsake throughout the Calgary Fire Department.

100 YEARS OF FIREFIGHTING

SMOKE, SWEAT, TEARS AND THE PEANUT BUTTER LEGACY

In 1985, the Calgary Fire Department celebrated its centennial with the publication of *100 Years of Smoke, Sweat and Tears*, a fond look back at the department's history.

THE BOOK PROJECT began in October 1982, when firefighter Bill Weisenburger started compiling photographs and anecdotes that could illuminate the Department's storied past, dating back to the volunteer brigade.

Over the next three years, Weisenburger and a volunteer team scoured archives and newspaper files for stories of bygone eras; they met with current and retired firefighters as well as family members to glean memories of life facing flames. The collection process was so successful that a companion book called *Milestones & Mementoes 1885 – 1985* was produced in a yearbook format. It was filled with pictures, newspaper clippings, personal tributes and a treasure trove of trivia including upkeep costs of horses, an annual record of major fires and a "nominal roll" listing members who joined the Department each year.

Weisenburger asked former Calgary Mayor Grant MacEwan, author of dozens of books on Western Canadian history and Alberta's Lieutenant-Governor for eight years, to write the anniversary book. MacEwan agreed and set his price: one jar of peanut butter. Best Foods Canada gave five cases of its Skippy peanut butter to the food bank in MacEwan's name following the book's publication. Best Foods donated another 1,000 jars on the author's 90th birthday in 1992.

▲ Project co-ordinator Bill Weisenburger showed off *Smoke, Sweat and Tears* as author Grant MacEwan was named honorary fire chief for his work on the book.

The novel payment was quickly adopted by other organizations and continued as a tradition until MacEwan's death in 2000.

"Over the years, many groups gave Grant peanut butter for his services," recalled Weisenburger. "At his 90th birthday, there was a party at the Palliser Hotel and there were cases of peanut butter tucked under the tables."

Out of the book project came a wonderful friendship. Weisenburger spent 15 years as an unofficial aide to MacEwan, assisting him with grocery shopping, driving to events and even modifying a set of headphones to improve the hearing of the aging statesman. Weisenburger was once described as "the son Grant MacEwan never had" and served as a pallbearer at his funeral.

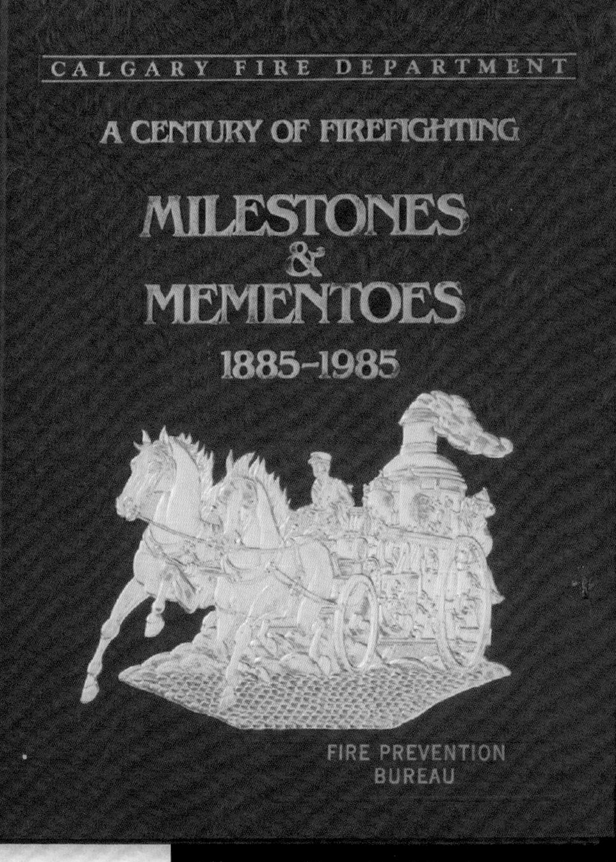

▲ Known as "the red book," well-worn copies of Milestones & Mementoes were regularly pulled out at fire stations as reference material.

The 5,000 copies of the centennial books were quickly snapped up, selling for $25 each and generating $35,000 in proceeds for the Firefighters Burn Treatment Society. *Milestones & Mementoes* sold out in six days. ■

▲ Mackenzie Nielsen, with Dalmatians Finn and Sally, was part of the popular 2009 *Hot Stuff* fundraising calendar.

RED-HOT FIREFIGHTERS AID BURN SOCIETY

▲ Many sons have followed their fathers into the Calgary Fire Department, but Jack and Kevin Collison were the first to share a pin-up page.

THE *HOT STUFF* CALENDAR was launched in 1986 to raise money for the Calgary Firefighters Burn Treatment Society. It was an immediate success, selling 15,000 copies the first year.

Filled with buff firefighters, the beefcake calendar originally featured firefighters involved with the popular Ladies Night Out events, another burn treatment fundraiser. Over the years, volunteers signed on to be the monthly pin-up.

One of the most unusual photos was the father-son combination of Captain Jack Collison and firefighter Kevin Collison, who shared the Mr. September title in the 1992 calendar. The Collisons were also Sunshine Boys in the Calgary Sun with the same image.

Each year, firefighters spend hundreds of off-duty hours raising money for highly specialized treatment and awareness of the Calgary Firefighters Burn Treatment Centre. The unit opened in 1987 at the Foothills Medical Centre with eight beds, providing care for burn victims from southern Alberta, Saskatchewan, British Columbia and northern Montana.

By 2008, the Society had raised $4.6 million for the burn treatment facility. ■

1989 THE UNDERWOOD BLOCK, BUILT IN 1911, WAS DESTROYED BY RAGING FIRE

CLASS OF 1989:

LOUISE VIAU-PICKERSGILL

▲ Viau-Pickersgill and an unidentified paramedic helped a child injured in a car accident.

A NEW FRONTIER

CALGARY'S FIRST FEMALE FIREFIGHTER

"I remember thinking 'If they can have jobs like that in space, I can be a firefighter.'"

Louise Viau-Pickersgill became one of Alberta's first paramedics in 1981 and worked out of Calgary fire stations for years before deciding to expand her first-responder skills. A tenacious weight lifter, Viau-Pickersgill wanted to use her strength for "more than just lifting patients."

JOINING THE CALGARY FIRE DEPARTMENT meant passing rigorous fitness requirements that included repeatedly hoisting a 35-kilogram coil of fire hose up a pulley and heaving those coils of hose onto a fire truck. After falling just short of the physical requirements in 1987, Viau-Pickersgill set up a testing ground in her backyard and spent hours each day training to qualify in 1989.

The Department made no special considerations for its first woman firefighter. That was just fine with Viau-Pickersgill, who wasn't interested in trailblazer status.

As a child, she was a devoted fan of the original Star Trek with its strong, intelligent female characters such as communications officer Lieutenant Uhura. Viau-Pickersgill was inspired by the portrayal of women as equals on the television series.

"I remember thinking 'If they can have jobs like that in space, I can be a firefighter,'" said Viau-Pickersgill, who retained her paramedic qualifications and was part of the heavy rescue team before becoming Assistant to the Department's Medical Director.

Fire Chief Jack Ross said the number of women applying for positions grew throughout the 1980s. But that could mean 20 applications out of the 2,500 people competing for 20 available jobs.

"Louise wasn't hand-picked," said Chief Ross. "We weren't trying to make ourselves an equal opportunity employer. We were one. Hiring Louise was a reflection of that."

Still, Ross remembered being impressed with her strength, agility and mind-set: "She demonstrated that never-give-up attitude."

Over the next several years, Viau-Pickersgill was asked to speak at various recruitment campaigns as the Department looked to expand its membership. But there was a conscious effort made to avoid drawing too much attention to her special status.

"You're not doing anyone a favour if you make them too unique," said Chief Ross.

The Calgary Fire Department had four female firefighters by the end of 1999: one on each platoon. ■

RESCUING "MAGGIE"

REGAINING THE PAST FOR THE MUSEUM SOCIETY

MORE THAN 50 YEARS after the Magirus aerial ladder truck arrived in Calgary heralded as an engineering marvel, the German rig returned as a museum showpiece.

In the late 1980s, Calgary Fire Department photographer Orlo Tveter spotted the 1929 fire truck, dubbed "Maggie," among a collection of old vehicles in Wetaskiwin, Alberta. It didn't have its original 85-foot wooden ladder and needed a little work, but the chassis was intact. "It was a classic piece that I thought Calgary should recoup," said Tveter, who mentioned his find to Training Officer Ray Ross.

What made the Magirus unusual was its aerial platform. While other aerial rigs needed to kick out stabilizer legs to lift the truck off the ground and provide support if the vehicle was on a hill or an angle, German engineers came up with the concept of leveling the platform instead, using a swivel. That was radical thinking in 1929 and it would take another 60 years before other manufacturers bought into the idea.

Ross and Tveter located a 100-foot Magirus ladder in Wisconsin and offered to sell the combination to the Firefighter Museum Society for their costs: $20,000.

Meanwhile, the German manufacturer was interested in regaining the aerial truck. Magirus offered to trade a new Mercedes Benz sports car for the aerial – a vehicle worth about $80,000 Cdn.

The pair rejected the tempting offer and brought the rig back to Calgary with one major caveat: if the truck was to ever be sold, Tveter and Ross would have first right of refusal to buy it back at their original price. The firefighters' union loaned the museum society the money to buy the rig. A special lottery was held among Calgary Fire Department staff to cover the loan and by 1991, the vehicle was safely tucked in the museum society's storage tent.

▲ Calgary's Class of 1946 posed on Maggie, one of only 13 Magirus aerials purchased by fire departments in Canada and the United States.

By 2009, Calgary's Magirus was one of two left in North America. "There are people who would come from across North America and Germany to see that rig alone because it's the only one operating," said Tveter. ■

LEADING BEHIND THE SCENES

"Few people ever knew the things he did for the community because he preferred to be anonymous."

A MEDIA-SHY OIL EXECUTIVE was named an honorary Fire Chief in 1996 for years of quietly helping build the stature of the Calgary Fire Department.

David Mitchell was the Chief Executive Officer of Alberta Energy Company, the forerunner of oil and gas giant EnCana, when he gave the Department $25,000 in 1993 to recreate the Cappy Smart Band. The group, formed by the former Chief in the late 19th century, folded in 1914 when all musicians in the band volunteered for service in the First World War.

"David Mitchell never, ever wanted accolades or to be seen on the front page," said Wayne Morris, Calgary Fire Chief from 1999 to 2005.

Mitchell was the first businessman to step up and sponsor Calgary's bid to host the World Police Fire Games, held in 1997. He would also work quietly behind the scenes, networking and helping to raise funds for public safety campaigns.

The family links to the Calgary Fire Department are meaningful. John (Jack) Mitchell, David's father, was a close personal friend of Cappy Smart and the two are resting next to each other in the Union Cemetery.

Luminaries of music and sport were also granted the title of honorary Fire Chief. Arthur Fiedler, longtime conductor of the Boston Pops Orchestra, was named an honorary Fire Chief when he visited Calgary in 1966, while golf legend Arnold Palmer received the Chief's helmet after donating his time to a fundraising golf tournament in the 1970s.

Former Calgary Mayor Grant MacEwan was named an honorary Chief in 1984 after writing a book celebrating the Department's first century of firefighting. ■

INDUSTRIAL ACCIDENT KILLS TWO

HUNDREDS EVACUATED AS BLAZE RAGES OUT OF CONTROL FOR ALMOST 10 HOURS

EXPLOSIONS ROCK SOUTHEAST CALGARY

PEOPLE LIVING NEAR THE HUB OIL REFINERY weren't initially frightened when southeast Calgary was rocked just before noon on August 9, 1999. After all, residents of Penbrooke Meadows were accustomed to smoke and fire coming from the Calgary Fire Department's nearby training academy.

But the erupting fireball that quickly filled the sunny summer sky made it apparent this was no training exercise.

Recruitment Officer Garth Rabel was screening applications and discussing lunch options when the training academy shook. Throwing open the blinds, he saw a massive mushrooming fireball rolling up and over the drive-in movie screens that lay between Hub Oil and the training centre.

Two employees at the oil recycling facility were killed when pressure buildup in an old storage tank caused the initial blast, destroying a building where the men were working. Firefighters from training were first on the scene as crews from several stations poured into the area.

As flames leapt from tank to tank, firefighters were driven back by two more major explosions that could be felt across Calgary. Waste lubricating oils, solvents and fuel within the vessels triggered up to a dozen smaller blasts. Pieces of debris were found embedded in homes and garages blocks away from the industrial site.

As the scenario intensified, emergency crews were forced to evacuate the area, abandoning an aerial ladder truck in the process. Oilfield firefighting experts Safety Boss, best known for extinguishing many of the fires in Kuwait set after the 1991 Gulf War, offered advice on quenching the blaze.

Hundreds of people in neighbouring communities were evacuated as the fire raged out of control for almost 10 hours. The incident raised questions about locating industry alongside residential areas and how stringently authorities were monitoring industrial facilities. ■

2000 TO 2010

CHALLENGES OF A NEW AGE INTO THE 21ST CENTURY

Calgary appeared to have it all as the 21st century dawned: a dizzying energy boom, massive population growth and a debt-free provincial future.

THE CITY'S POPULATION hit one million in July 2006. The milestone followed a decade where more than 100,000 people moved to Calgary from other parts of Canada, seeking a slice of an economic pie that was the envy of the rest of the country.

As the decade progressed, recruitment became a huge issue for the Calgary Fire Department. Retiring firefighters would need replacements and getting qualified candidates was difficult in a rapidly expanding city where Help Wanted signs were everywhere. Late in 2006, an aggressive campaign was launched to attract physically fit men and women – drawing 267 recruits in 2007 and 2008. Yet with 40 per cent of Calgary firefighters expected to retire by 2010, demands for superior recruits would continue.

A major aid to prepare those new firefighters became available in 2004 when the Multi-Agency Training (MAT) Centre opened its doors at the department's Training Academy in Forest Lawn. The MAT Centre was also home to Canada Task Force 2, an urban search and rescue team that trains emergency responders and volunteers for natural disasters or other catastrophic events. Meanwhile, the Fire Department was stretched to its limits. Years of holding the line without adding to the service were starting to show – especially in the burgeoning suburbs.

In 2006, City Council committed $293 million to improve emergency response. Much of that investment would go towards replacing the aging Fire Department fleet with 17 new vehicles. It would also help build four new fire stations, an Emergency Operations Centre and a new vehicle maintenance shop.

All new facilities would adhere to the City of Calgary's stringent environmental guidelines aimed at reducing the impact on the planet. The Fire Department was on the leading edge of the green movement, retrofitting a truck in 2004 that would run on biodiesel, a combination of petroleum product and oils extracted from canola and other grains.

◄ In 2009, Calgary firefighters received new hat badges with regimental badge numbers. The pin is capped with a ridge of mountains and includes the city skyline. Several badge numbers were retired in honour of those who died in the line of duty.

INGENUITY REMAINED AN INTEGRAL PART OF THE FIREFIGHTER TOOLKIT

In 2008, fire personnel custom-designed a cutting-edge Hazardous Materials Response Vehicle to clean up hazardous and flammable materials from city roads and properties.

With little warning, the global economy crashed as 2008 drew to a close.

In a six-month stretch, oil plunged more than $100 a barrel, sending shock waves and layoff notices through Calgary's oil and gas towers. Major energy projects were shelved, construction slowed dramatically and by early 2009, hundreds of thousands of people across Canada lost their jobs. Calgary was in better shape to cope than most communities, but steady paycheques were no longer guaranteed.

In a sea of bleak economic news releases, Fire Chief Bruce Burrell announced his Department would hire 200 recruits in 2009. Potential recruits poured in the doors. ■

THE POWER OF MOTHER NATURE

PINE LAKE TORNADO

THIRTY SECONDS OF DESTRUCTION in a central Alberta campsite: that's all it took for the Pine Lake tornado to kill 12 people, injure hundreds more and spawn countless nightmares.

For Calgary firefighters, Pine Lake was a whole new rescue experience.

From the moment the deadly funnel cloud touched down southeast of Red Deer at dinnertime July 14, 2000, it was apparent the region's emergency resources would need reinforcements.

The twister was heading toward the waters of Pine Lake when its swirling 300-kilometre an hour winds ripped apart the Green Acres campground. Dozens of mobile homes and boats were crumpled and torn apart before being dumped in the lake.

Ninety minutes away in Calgary, Training Officer Keal Prince was alerted that rescue support was vital. Emergency crews feared the death toll could easily mount.

"In our minds there was hope of finding someone in a capsized boat with an air pocket," said Prince, an aquatic rescue expert with the Calgary Fire Department.

Teams arriving at the disaster scene were stunned by the scope of the devastation. Mattresses and vehicle parts skewered onto trees. Mangled remains of recreational vehicles alongside a pristine picnic table set for supper, untouched by the storm's fury. Two trees, the smallest with a trunk the diameter of a dinner plate, fused into a makeshift cross.

Nothing in firefighter training had prepared them for this.

"We were walking into a war zone," said Prince. "Where do you start?"

Firefighters fanned out in ankle-deep water and slowly moved through a meticulous grid pattern, feeling through wires, broken glass and capsized vehicles. The stench of raw sewage from trailer septic tanks was heavy in the air as they inched along, eventually reaching water up to their shoulders. All were later given tetanus shots to ward off infection.

Over the next four days, Calgary's aquatic and heavy rescue specialists helped with the painstaking recovery mission.

They found no miracle survivors. Despite earlier fears, no bodies were discovered in the lake.

"We found a couple of dead ducks and a dead cat," said Dan Frederick, who was dive team leader at the time. "As the days wore on, everyone seemed to be accounted for."

"Rescue is utopia, it's the ultimate," said Prince. "But if you find the body of a loved one, you've provided closure for a family member. By showing up the first night, CFD was there to help. It's not that we were looking for Calgarians, we were looking for people." ■

2000 Military divers dragged high-tech side-scanning equipment over the murky waters, targeting places and objects for fire crews to examine more closely.

"WHITE POWDER CALLS"

soared across North America following the 2001 terrorist attacks, as panic spread far beyond the 9/11 sites. Fire's hazmat specialists were repeatedly called to Calgary office towers to deal with suspicious packages after letters poisoned with anthrax killed five people in the United States. By the end of the decade, Calgary's hazardous materials team was still receiving about one white powder call a month.

CLASS OF 1981: BRIAN BASWICK

HONOURING THE ETHIC

▲ Brian Baswick drank milkshakes for years to put on pounds, but was considered too light for the Calgary Fire Department. As soon as the weight restrictions were lifted, he was "in there like a dirty shirt" recalled his widow.

◄ James Baswick received the award named for his father during recruit graduation ceremonies April 7, 2006 from stepmother Wendy Baswick.

BRIAN BASWICK never looked like the stereotypical firefighter, yet he exemplified the best of the breed.

After being diagnosed with cancer, Baswick continued to train the next generation of firefighters until shortly before his death in 2001 at age 53. His memory lives on in the Baz Award, which recognizes the recruit chosen by other potential firefighters as the best all-around team player.

Slightly built and wiry, Baswick joined the Department in 1981 and became a Training Officer in 1988. He quickly developed a reputation as a caring instructor, thoughtful, obliging and conscientious.

"He was a great guy," said Community Safety Co-ordinator Garth Rabel, who helped draft criteria for the Baz Award. "His knowledge and ability to share that knowledge made him bigger than life."

The trophy, first given out in 2001, depicts a firefighter supporting a struggling colleague as the pair move away from a scene.

"The idea was the guy who would get the award was the one you wanted on your team," said former Training Officer Bruce Kinnell. "Our whole training is teamwork. We always stress the things you do together: you're not going to go out to a fire by yourself, you're never going to get a ladder by yourself."

Wendy Baswick, Brian Baswick's widow, hands out the award at each graduation ceremony. The presentation is often emotional, but it was especially poignant when Mrs. Baswick gave the honour to Brian's son James, who joined the Calgary Fire Department in 2006. ■

THE DUCTMAN OF DRUMHELLER

CALGARY MAY 18, 2002 ■ 96 PAGES, 2 SECTIONS ■ Vol. 22, No. 186

www.calgarysun.com

THE SATURDAY SUN

SUPER Have YOU won? winning numbers on page 2

Members of the Calgary Fire Department's 9 Heavy Rescue unit, who found missing Drumheller Institution convict Raymond Tudor, below, after he spent seven weeks hiding inside the penitentiary. In front are Mark Turik, left, and Jeff Trudeau. Behind them are, left to right, Randy McRae, Trevor Kerr, Captain George Lattery and Shane Lacharite. The crew is holding the tools they used to find and extricate Tudor.

How did this man elude officials for so long inside prison walls? You won't believe it!

—MIKE DREW, Calgary Sun

JAILHOUSE SHOCK

Cunning con finally caught after hiding in prison for 7 weeks: P. 4-5

The Calgary Fire Department's heavy rescue team made international headlines in 2002 for their seismic snooping.

THE SPECIALTY CREW used Delsar seismic listening equipment to locate a killer convict who spent seven weeks above the prison workshop at Drumheller Penitentiary northeast of Calgary.

"We thought he was stuck in the ventilation system," said Team Instructor Mark Turik. "We expected we were going to a rescue, not that he was hiding."

Raymond Tudor had a reputation for blending into the woodwork. After murdering two elderly victims in the 1990s, he concealed himself in the attic above the crime scene for days to watch investigators. Privately, police referred to him as a human chameleon.

Tudor disappeared from the penitentiary March 26, 2002, but no public alarms were sounded because prison officials weren't sure he had left the facility. Weeks of searching turned up nothing until a man resembling the missing inmate was spotted in the workshop after hours. That triggered a three-day lockdown as guards unsuccessfully combed through the complex.

The warden appealed to the Calgary Fire Department for help after hearing the heavy rescue unit had gear for locating people in tight spaces.

Once arriving at the prison, firefighter Trevor Kerr clipped his high-tech listening device into air ducts above the workshop. Almost immediately, Kerr heard what sounded like someone rolling over. An RCMP police dog verified the discovery, prompting the rescue team to cut a hole in the duct system and insert a snake-eye camera.

Tudor lay there motionless. RCMP and firefighters peeled back the sheet metal using a device similar to a giant can opener and uncovered the fugitive in his "nest," a small chamber cut into the narrow space between the ductwork and the roof. He had built a maze with false walls, trap doors and access holes, surviving on food smuggled in by other inmates.

Calgary firefighters were lauded for the unorthodox capture. ■

◀ Calgary's heavy rescue crew gained headlines around the world for helping to capture a missing killer.

ERLTON

"A MASTERFUL PIECE OF FIREFIGHTING"

A rookie roofer's mistake with a propane torch sparked Calgary's largest residential disaster – $66 million of destruction at an upscale condominium complex.

WHAT BECAME KNOWN AS THE ERLTON FIRE began at a construction site just south of downtown and spread like wildfire through occupied condos May 30, 2002. The blaze threatened city landmarks like the Stampede Grounds and the Talisman Centre athletic facility. Amazingly, there were no fatalities or civilian injuries.

Captain Neal Dennis vividly recalled the speed of the fire. Dennis was the initial fire ground commander and within minutes of his arrival, six buildings were in flames.

A newspaper image of Dennis running backwards doesn't show the portable radios in each hand as he directed crews into place while communicating with dispatch operators. Behind Dennis, 250 condo residents were yelling and pointing at fires breaking out on rooftops throughout the area.

▲ Firefighters battling the Erlton fire also had to contend with downed electrical wires.

◀ Captain Neal Dennis scrambled to move fire crews into place as the Erlton fire jumped from one building to another.

Rigs responding to four full alarms within the first 34 minutes pulled in at a pace that surprised even the veteran firefighter. Some appeared without being called.

"They were arriving bang, bang, bang," said Dennis. "We had an abundance of manpower there shortly."

At one point, all but five of Calgary's fire vehicles were battling the inferno as 140 fire personnel focused more than 50 pieces of equipment on the attack.

As an unexpected bonus, equipment and crews normally stationed in other parts of the city were downtown getting identification photographs for the G8 meeting of world leaders set for later in the summer.

The blistering heat of the fire was complicated by unseasonably warm weather and temperatures above 30°C. Two blocks away from the core of the blaze, an electronic billboard registered a scorching 46°C.

Dozens of firefighters were treated for heat exhaustion and minor injuries suffered while fighting the fire. More than 30 police officers had to be treated for smoke inhalation as they evacuated nearby residents.

But heat and smoke were the least of the dangers facing fire crews.

"The main transmission [power] line was arcing and sparking and looked like the 4th of July," said East District Chief Nick Maley, who took over site command.

Live power lines were scattered on the streets. Explosions came fast and furious. A lobby erupted without warning, either from a gas buildup or backdraft, sending its metal-framed double door rocketing past where fire crews had been set up just seconds before. Missile-like debris damaged fire trucks instead of taking lives. The force of the blast embedded chunks of glass in fences across the street.

"The fire was creating its own windstorm," said Maley. "If the fire got away, we would have been calling it a disaster area."

Embers and ash ignited dozens of smaller fires.

"I was overwhelmed by how hard each individual worked and how well their training kicked in so that we could manage the incident," said Maley, who retired in 2007.

"Given the extreme conditions and the challenges faced by incident commanders and fire crews, this was a masterful piece of firefighting," said Deputy Chief Len MacCharles. ■

2002

▲ Perry Andersen, described by other firefighters as possessing "a work ethic that makes a rookie look bad," sat exhausted after going through several tanks of air searching condo suites for people or pets.

◀ Fire swept through the upscale condo development like wildfire, ultimately causing $66 million in damage.

▲ Larry Sorby held a partially burned rubber boot while firefighters searched for colleague Jerry Walter, who was killed fighting a restaurant fire in April 1972.

DUTY WITH HONOUR

THE ULTIMATE SACRIFICE

▲ During the memorial's 2006 dedication, the bell tolled once for each firefighter who died in the call of duty.

EACH SEPTEMBER, a formal service is held in front of City Hall honouring Calgary firefighters killed in the line of duty. A special tribute monument was built in 2005 to pay respects to the city's firefighters and police officers who have given their lives in protection of their fellow citizens.

Historic photographs of fire and police are etched on outside walls of the black granite square. Images of those who died are carved on inside walls.

The Calgary Fire Department Honour Guard wears a special patch on its dress uniform with nine red stars: one for each firefighter who made the ultimate sacrifice. ■

LAST ALARM

FIREFIGHTER
Hugh McShane
• Nov. 17, 1923 (fire vehicle accident)

CAPTAIN
Arthur Simmons
• Dec. 30, 1948 (Union Packing Plant fire)

FIREFIGHTER
Norman Cocks
• May 8, 1962 (training accident)

LIEUTENANT
Lloyd Dutnall
• Sept. 6, 1970 (Calgary Stockyards fire)

LIEUTENANT
Harold E. Smith
• May 27, 1971 (McTavish Block fire)

FIREFIGHTER
Jerry Walter
• Apr. 20, 1972 (Beachcomber Nightclub fire)

FIREFIGHTER
David F. Allan
• Aug. 25, 1976 (fire vehicle accident)

FIREFIGHTER
George Look
• Jan. 1, 1981 (Manchester Racquet Club fire)

FIREFIGHTER
John Morley James
• July 13, 1992 (Forest Lawn Hotel fire)

In April 2003, Alberta became the second province to recognize in law that fighting fires puts professional firefighters at a higher risk to develop certain cancers. Tribute is also paid to firefighters whose deaths were ruled work-related by the Workers' Compensation Board:

CAPTAIN (R)
Ron Renard
• July 2, 1998

DIVISION CHIEF (R)
Ken Moody
• Sept. 14, 1999

CHIEF TRAINING OFFICER (R)
George C. Heming
• June 26, 2003

CAPTAIN (R)
Edward E. Briggs
• Jan. 23, 2004

DEPUTY CHIEF (R)
William (Bill) H.G. Beattie
• May 9, 2005

CAPTAIN (R)
Alan H. Edwards
• May 12, 2005

FIRE CHIEF (R)
Thomas E. Minhinnett
• June 28, 2005

DISTRICT CHIEF (R)
Reuben Poffenroth
• Sept. 5, 2005

CAPTAIN (R)
James M. Symon
• Nov. 24, 2005

FIREFIGHTER (R)
Robert N. Elder
• Sept. 9, 2005

CAPTAIN (R)
Bruce B. Dancy
• Aug. 31, 2006

DISTRICT CHIEF (R)
Sidney B. Gilbert
• March 15, 2007

FIREFIGHTER (A)
Gord Paul
• Aug. 7, 2007

CAPTAIN (R)
Olaf Wilson
• Oct. 17, 2007

DIVISION CHIEF (R)
Jim Carrington
• Nov. 24, 2008

(A) – Active Member
(R) – Retired Member

CLASS OF 2002: RANDY SCHMITZ

SAFETY BUSINESS RISKIER

SAFETY SELLS in the car market, but few salespeople mention those features that protect passengers in a collision can impede any rescue.

"They don't make rescuer-friendly vehicles anymore," said Randy Schmitz, the Calgary Fire Department's vehicle extrication instructor. "Devices put in to protect passengers (can) kick the crap out of us."

Cutting victims out of crumpled cars is a skill requiring constant upgrades and Schmitz is one of the few people who have made it a specialty. He began developing vehicle extrication expertise in 1992 and teaches fire departments across North America, including annual updates for Calgary firefighters.

Schmitz has seen a jump in accidents caused by cellphone use and says people also have a tendency to drive faster if they believe their vehicles can protect them in a crash.

And manufacturers have created good reason for those comfort levels.

Increasingly tougher frames are being crafted to ensure safer vehicles, while multiple inflation devices have become the norm. By 2009, new vehicles needed a minimum of six airbags. Any that don't inflate in a collision are hazards, with cylinders of pressurized gas becoming potential bombs if cut by a rescue tool.

In 2007, Calgary invested $1 million in new extrication equipment to ensure faster rescue of people trapped in vehicles following collisions. But keeping pace with improved safety measures is a challenge.

Disconnecting batteries, which stops electrical power from surging to potential hazards, is also more complicated. A growing number of vehicles don't have the battery under the hood.

All of these safety improvements increase rescue times.

Schmitz says it should leave drivers with a sobering thought: "There could come a time when you can't be rescued from a super safety vehicle." ■

Y-FIRES

PLAYING WITH FIRE NO GAME

IN THE 21ST CENTURY, fire prevention officers moved away from taking young Emma to burn treatment units for a fire consequence lesson to involving her in an individual plan to douse dangerous behaviour.

The Calgary Fire Department developed the Youth Fire setter Intervention Referral and Education (Y-FIRES) program to connect with kids before natural curiosity escalates into major destruction. Through confidential home visits, specially trained firefighters work with juvenile fire setters and their parents to stop their actions.

Studies indicate that four out of five kids who don't get help will set more fires.

More than 360 youths went through Y-FIRES between 2004 and 2009, said Dean Krupa, Community Safety Officer in charge of the program. Only two who completed the sessions were known to set more fires.

"There are probably at least another 500 we don't hear about," said Krupa. "If Johnny lights grass fires behind the school shed and gets suspended for two days, we're probably not going to hear that." ■

2006
DREAM COMES TRUE FOR GOLDEN OLDIE

FIREFIGHTER DUFF GIBSON, 39, became the oldest Olympic gold medallist in Winter Games history when he won the men's skeleton event in 2006. Gibson's crewmates at the Calgary International Airport fire station welcomed him home from Turin, Italy with a parade of flashing fire trucks that escorted the aircraft carrying Canadian athletes into the terminal. Gibson retired from competition but continued his interest in skeleton, helping to coach athletes for the 2010 Games in Vancouver during his vacation and time off from firefighting. ■

RIVER RESCUE

Jeff Budai, a member of the dive team, pulled a man to shore near the Calgary Zoo after the man's car plunged into the Bow River in May 2006.

2008 **FIREFIGHTERS DEMONSTRATE** putting an oxygen mask on a dog after the Calgary Fire Department began using special CPR equipment and techniques for pets in distress. The equipment, carried on all fire pumps and rescue boats since 2008, includes oxygen masks designed for various sizes of dogs and cats.

CLASS OF 1934: ROY STEADMAN

CALGARY'S 100 YEAR OLD FIREFIGHTER

WIND IN HIS HAIR, sleet on his face. Roy Steadman's firefighting memories are all tinged by weather conditions.

"I never had a closed truck," says the retired District Fire Chief, who turned 100 on November 5, 2008. "They may have been around when I retired, but the only time I remember being in any closed truck was when I was a District Chief."

Generations of Calgary firefighters spent their entire careers riding on the back of a rig: rain and snow blasting into their faces as they rushed to a fire. Steadman was among the class of 1934, a group of 20 young men hired to cover shifts when the City of Calgary granted firefighters one day off in seven.

"The firefighters have some nice equipment now," Steadman says in a wistful voice. "Big pumps: rigs where the guys can ride inside, not outside. That's a big improvement."

Few people can look back over an entire century of change, but the feat is almost unheard of for firefighters.

Steadman retired in 1969 after 35 years with the Department, the first in his class to turn in his helmet. By the time he reached the 100-year milestone, he had been gone for longer than he was part of a crew.

Still, some memories never fade away. Like his first bad fire.

It was a cold winter morning in 1937 in the hours before sunrise. A father had gone to work while his three young children slept off the kitchen and their mother dozed in a nearby bedroom. As the father hiked to work, he took little notice of the fire truck that raced past him, unaware it was heading for his home.

Inside the four-room cottage, firefighters found a little girl in a cot, blankets pulled up over her head. Her two younger brothers were curled up together in a crib. Even the family dog was dead. All were overcome by fumes from an improperly closed coal-fired stove.

"She was such a pretty little thing, with a head of long blonde curls," says Steadman, who had two of his own daughters. "I'll never forget her."

The children's mother was the lone survivor, spared only because her bedroom window was partially ajar.

Working at the Fire Department was Steadman's first steady job and the only one he ever really wanted. It allowed him to marry his longtime sweetheart, Edna, and raise a family. Over the years he worked on the Department's annual toy campaign for needy children and after retiring, headed the Pensioners' Association for 16 years.

"It's been a good life," he says with a smile. "And it was a good job." ■

Roy Steadman died October 9, 2009 – just shy of his 101st birthday.

COMMUNITY CONNECTIONS

TOUCHING HEARTS AND TICKLING FUNNYBONES

▲ In 1967, Calgary firefighters began hosting a giant party for needy children and their families every December. The clown team is a popular fixture at the annual bash.

► For CFD mascot Sparky and friends, clowning around at the kids Christmas party is an important way to kick off the holiday season.

TO MANY PEOPLE who have never needed emergency help, Calgary's firefighters are community heroes because of their volunteer work. Whether counting the coins in UNICEF boxes after Halloween, playing bagpipes to honour a fallen comrade or wrapping toys for underprivileged kids, the off-duty activities of firefighters make a major contribution to Calgary's heart.

The Calgary Firefighters Toy Association began in the 1940s with members restoring used toys to make Christmas brighter for children whose families were struggling. More than 60 years later, the toy repair shop is closed but the tradition continues with an annual party at the Stampede Corral the third weekend of December.

About 1,500 deserving families are invited each year, referred by school officials, the Salvation Army and firefighters themselves. Santa arrives in an antique fire truck and all children under 12 are given a new toy. Each family takes home a Christmas tree.

Active and retired firefighters dedicate hundreds of volunteer hours to ensuring the success of the $50,000 annual event. The Calgary Firefighters Clown Team delights the crowd, happy to stand in for Santa's elves in spreading the holiday spirit.

"I've been to calls after this party where all they had (for Christmas) was one of our trees and a couple of our presents," said Gord Robb, vice-president of the 2009 Toy Association.

Firefighters have distributed more than 100,000 toys to Calgary children at Christmas. The toy association is also able to help families at other times of the year. ■

MANY FIRE SERVICES ACROSS NORTH AMERICA established or revived pipe bands following the deaths of 343 New York firefighters in the 9/11 terrorist attacks. Since 2002, Calgary's Pipes and Drums band has represented the department at dozens of events each year, including Fire Training Academy graduations, the Stampede parade and memorials for line-of-duty deaths within the fire community.

SETTING THE STANDARD

FIREFIGHTING IN THE FUTURE

Fire Chief Bruce Burrell envisions a day when all eyes look to Calgary's fire service for leadership and advice.

"**IF YOU WANT TO KNOW THE BEST WAY TO DO IT,** pick up the phone and call Calgary," said Chief Burrell, a Halifax native who took over the Calgary Fire Department's top job in 2005.

"We want to be the leader not just in one facet, but in every facet of what we do."

Chief Burrell wants Calgary recognized as the best fire service in the world. Part of that process is confirming the Department's certified status with the Commission on Fire Accreditation International. In 1999, the Calgary Fire Department joined an elite group by meeting the commission's high-efficiency standards – the first Canadian fire service to receive the prestigious designation. The Calgary Fire Department was successful in reaccreditation for 2010-2015 in late 2009.

Continuing to build on these achievements will help Calgary evolve from a fire-based organization into a true public safety organization.

Chief Burrell wants the Fire Department to reduce its environmental footprint and to have more connection with the community. It may seem difficult to improve on the Department's public approval rating of 98 per cent, but that's the plan.

"Customer service focus and value added programs will remain critical to the Department's sucess with one in eleven Calgarians accessing those programs in 2009," said the Chief.

► Chief Bruce Burrell wants building codes improved to slow the spread of flames in house fires, such as this 2006 blaze in Tuscany that caused $1.2 million damage to five homes and left three families without shelter.

The Fire Department assumed a national leadership role when Chief Burrell became President of the Canadian Association of Fire Chiefs in 2008. That meant Calgary's calls for building-code changes would be heard by more people. Some construction materials are more flammable than other products and new homes are being built closer together than in the past. The combination means a fire that would destroy a single house can now consume three in the same amount of time. Working with other fire chiefs and the National Research Council, Chief Burrell's goal is to see safer houses built that would also be less dangerous for firefighters and the public in case of fire.

Back at home, he wants to establish an office of innovation and build on Calgary's trailblazing reputation that dates back to Cappy Smart. Firefighters and civilian staff would be encouraged to submit ideas for change. Those suggestions will be evaluated and the best ideas brought forward for implementation.

"We want to continue to be on the leading edge," said Chief Burrell, adding that creativity and success build an infectious synergy other departments will want to emulate. ■

VEHICLES USED BY THE CALGARY FIRE DEPARTMENT IN 2010

Aerial No. 18

Engine No. 35

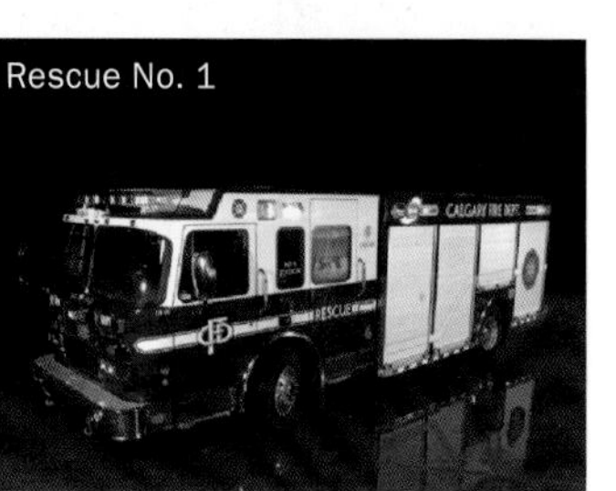
Rescue No. 1

Bronto No. 12

Hazardous Materials Vehicle

INDEX

FIREFIGHTERS (BY NAME)

PHOTO AND ILLUSTRATION CREDITS

YOURS FOR LIFE: 125 YEARS OF COURAGE, COMPASSION AND SERVICE FROM THE CALGARY FIRE DEPARTMENT

Printed in Canada.

Library and Archives Canada Cataloguing in Publication

Yours for life : 125 years of courage, compassion and service from the Calgary Fire Department.

Issued by: Calgary Fire Dept.

ISBN 978-0-9685795-4-1

1. Calgary (Alta.) Fire Dept.–History.

2. Fire prevention–Alberta–Calgary–History.
I. Calgary (Alta.). Fire Dept
II. Title: 125 years of courage, compassion and service from the Calgary Fire Department.

TH9507.C34Y58 2009 363.37'809712338 C2009-906444-8

UBLIC MARKET